D0209480

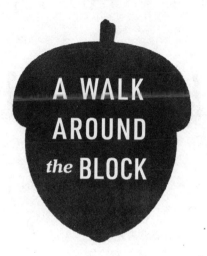

A WALK
AROUND
the BLOCK

A WALK AROUND

the BLOCK

Stoplight Secrets,
Mischievous Squirrels,
Manhole Mysteries
& Other Stuff You See Every Day

(and Know Nothing About)

SPIKE CARLSEN

HarperOne
An Imprint of HarperCollins*Publishers*

A WALK AROUND THE BLOCK. Copyright © 2020 by Spike Carlsen. All rights reserved. Printed in the United States of America. No part of this book may be used or reproduced in any manner whatsoever without written permission except in the case of brief quotations embodied in critical articles and reviews. For information, address HarperCollins Publishers, 195 Broadway, New York, NY 10007.

HarperCollins books may be purchased for educational, business, or sales promotional use. For information, please email the Special Markets Department at SPsales@harpercollins.com.

FIRST EDITION

Designed by Janet Evans-Scanlon

All photographs courtesy of the author unless otherwise noted.

Squirrel image, pages i and ii: Potapov Alexander | Shutterstock

Library of Congress Cataloging-in-Publication Data

Names: Carlsen, Spike, 1952– author.
Title: A walk around the block : stoplight secrets, mischievous squirrels, manhole mysteries & other stuff you see every day (and know nothing about) / Spike Carlsen.
Description: New York : HarperOne, 2020 | Includes bibliographical references.
Identifiers: LCCN 2020025829 (print) | LCCN 2020025830 (ebook) | ISBN 9780062954756 (Hardcover) | ISBN 9780062954763 (Trade Paperback) | ISBN 9780062954770 (eBook)
Subjects: LCSH: Curiosities and wonders. | Material culture—Miscellanea. | Recreation—Miscellanea. | Inventions—Miscellanea. | Questions and answers.
Classification: LCC AG243 .C358 2020 (print) | LCC AG243 (ebook) | DDC 031/.02—dc23
LC record available at https://lccn.loc.gov/2020025829
LC ebook record available at https://lccn.loc.gov/2020025830

20 21 22 23 24   LSC   10 9 8 7 6 5 4 3 2 1

For Kat; now it's our turn to walk

· CONTENTS ·

PART III: **SURFACES**

PART IV: **NATURE**

PART V: SIGNS, LINES, AND LIGHTS

· INTRODUCTION ·

WHY AM I—ILLITERATE IN FRENCH AND IGNORANT OF LAW— crouched beside a stranger, spray-painting my name on a Paris alley wall? Why am I stumbling around a two-hundred-year-old sewer exploring sludge and rats? Why am I interviewing a squirrel linguist? Surely there are more legal, less olfactorily offensive ways to find out about the world outside my front door.

I can only blame it on one bitterly cold morning a few years back. I had shuffled into the bathroom and turned on the water to brush my teeth. Nada. In a fog, I trudged to the kitchen faucet hoping for better. Nope. My next shuffle was to my phone to call my city water department. The voice that answered was that of someone who'd already answered the phone one too many times that morning. Halfway through my first sentence, the voice sighed. "I'll put you through to Robert."

"This is the fifth call we've gotten from people in your part of town this morning," Robert said. "Your water service line is frozen."

This was impossible. This is America. This is the twenty-first century. This is "The Land of 10,000 Lakes."

Robert explained there was only one company that could remedy the problem, and it was booked for two days.

This too seemed impossible.

Two days later, a sleep-deprived worker from Miller Excavating showed up in a flatbed truck with a massive gas-powered arc welder crouched on the bed. He snaked one end of a long cable down the basement stairs and attached it to a pipe coming out of the floor. Then he went out to our driveway, kicked around in the snow until he found a hockey puck–size disk, and attached another cable. He told me to turn on our kitchen faucet, then fired up the arc welder and a Marlboro Light. He explained the welder was feeding current through one cable, using the metal water line to conduct current to the other cable, thus generating heat and—hopefully—thawing the water line. Two cigarettes later, the faucet burped and a string, then a rope, then a flush of water emerged. Then I wrote a check for $250.

I called Robert at the water department to give him the all clear. "Unless you wanna go through that again," he sighed, "open a faucet and let a pencil-lead stream of water run for the rest of the winter." It was mid-January. Surely he meant rest of the DAY. No. For the next six weeks, we let a pencil-lead stream of water run into the sink.

The endless trickle became a gnawing reminder of everything I didn't know. I really had no idea where our water came from. Or where it went. Or, for that matter, how my phone call to Robert had gotten to Robert. As I stared out the frost-etched window, I further realized I knew nothing about the concrete sidewalk leading to our door, the front lawn beneath the freshly fallen snow, the squirrels in the trees, the ancient walnut tree on the boulevard, the graffiti on the back of my neighbor's garage . . . nothing. And these were just the things I could see through one dinky window. I realized I'd read books about journeying across Antarctica, down the Amazon River, and up Mount Everest; I'd written books about fifty-thousand-year-old wood buried in the bogs of New Zealand

and the violin makers of Cremona, Italy—but I knew nothing about the world right outside my front door.

A curious writer should do something about that.

When I launched into the research, I envisioned sitting at the Daily Grind, coffee steaming, fingers tapping, search engine searching. But the more I researched, the less I sat. I found myself hanging out with pigeon racers, traipsing through scorching power plants, stumbling through recycling facilities, and walking, stride-for-stride, with the Nordic Walking Queen. I meandered through Oakland Cemetery in Atlanta in search of the gravesite of the author of *Gone with the Wind* and through the streets of Carmel, California, to secure a copy of a permit for wearing high heels. It turns out that the story of things isn't so much about things as it is about people—their triumphs, failures, obsessions, and brilliance. It's about history, myths, and the future. It's about how things affect us and how we affect things. Those in academic and bureaucratic circles refer to these "things"—these pipes, wires, roads, signs, systems—by the soda-cracker-dry term "infrastructure." I like to refer to them as "the things that sustain us," as the awesome, essential, hidden and not-so-hidden world around, above, and below our feet.

What will you get out of this book? My guess is you'll never look at a stoplight, squirrel, or manhole the same way. Just as knowing the rules, quirks, and history of, say, baseball or ballroom dancing makes watching these affairs more exciting, knowing the inner workings of the world outside your front door makes life more interesting. At a minimum, you'll be a brilliant conversationalist at parties and while stuck in elevators. But there's more. The section on walking may help you live seven years longer. The section on trees may help you, your house, and your planet remain cooler and calmer. The interview with a founder of the zero-waste movement may radically change the size of your garbage can. The section on stoplights could save your, or someone else's, life. This awesome world isn't just a spectator sport. It's symbiotic; it influences us, and we influence it. We have some skin in this game.

Scientists have calculated the impact it would have if the entire population—all eight billion people—gathered in one spot and jumped at the same time. That impact would shift the earth's orbit the mere width of a hydrogen atom. But, if even a fraction of those people filled a pothole, installed a solar panel, or biked a little more, the impact would be real. Here you'll discover a little more about what's worth jumping about.

Novelist Milan Kundera said, "To be a writer does not mean to preach a truth, it means to discover a truth." What better way to discover truth than with nose plugs on and a can of spray paint in one hand? What better way to explore than together? Let's go discover some stuff, a lot of stuff. Come on—let's take a walk around the block.

PART I

INCOMING

1

THE FRONT PORCH

Rocking Back *and* Forth
with the American Dream

WHEN JOHN AND CAROLINE PROCTOR PAID THE PRINCELY SUM OF $2,600 in 1850 to have the Greek Revival–style house my family and I now live in built, they made sure the architect included a front porch—not a large affair, one just big enough for two rocking chairs and a small table.

John Proctor wore a variety of hats. He owned a general store offering "lynx muffs, buffalo overshoes, and fresh butter." He served as prison warden, instituting time off for good behavior and the mandate that prisoners wear striped uniforms, which made escapees easier to spot. Proctor was elected to three terms as mayor and was appointed by Governor Pillsbury as "surveyor general of logs and lumber."[1]

Often, I enjoy imagining the prosperous couple sitting on that porch, rocking and chatting. From their overlook, they could watch the massive booms of white pine float down the St. Croix River. On hot summer days, the shade would have offered respite from the heat. But I'm guessing more than anything, the front porch served as an informal backdrop for holding court with citizens as they strode by.

Over the years, as the layout of the neighborhood changed, the front porch became the side porch. The porch still offers relief from the heat and a view of the river, which now teems with restored paddlewheel boats. But since it no longer faces the street, it no longer provides that informal setting for "conversation, news, gossip, bickering, courting, entertaining, and community affairs . . . a place for neighbors to be neighbors" that it once did.[2]

Porches also made functional sense. Whether you were a caveman dragging home a wildebeest or an accountant dragging home a bag of groceries, a sheltered place to pause before entering your abode made all the sense in the world. In scorching climates, covered entryways became logical extensions of the house; in Africa, a grass roof supported by poles provided shade, the openness provided ventilation, and a raised floor offered respite from snakes and biting insects.

As different cultures settled in the United States, they brought their architectural language with them. Creolization—the blending of African, Caribbean, and European cultures—played a large role in the development of porches in the Deep South. The simple front porches of shotgun houses were most likely a carryover from African and Haitian architecture. The Dutch brought with them the concept of a raised first floor and front stoop, built to elevate their houses above the omnipresent threat in their homeland of flooding.

Yet there's something distinctly American about the American front porch. It wasn't just a physical object; it showcased social and political issues as well. Colonists after the Revolutionary War wanted nothing to do with things that whiffed of English dominance, so a new American architecture arose. Every respectable house of the Victorian, Queen Anne, Greek Revival, or Craftsman era was built with a confident porch; houses on corner lots often had two.

The golden age of porches glowed its brightest between 1870 and 1920. In an era when the backyard often contained features of drudgery—vegetable gardens, trash pits, outhouses, perhaps clucking chickens or a goat—the front porch offered an oasis of calm.

In 1888, John and Caroline Proctor had a second house built for them three blocks to the south. By then they'd sold enough lynx muffs and buffalo overshoes to afford something grander. They didn't have to tell the architect to include a front porch, since *not* including one would have been architectural blasphemy. That second Proctor house has a grand front porch spanning the entire length of the house. Six of our porches would fit on that one. It has stone supports, ionic columns, and fancy moldings.

Coincidentally, two of our best friends, Paul and Laura, purchased the house five years ago; we've spent plenty of evenings there, sitting and sipping. A block to the west live our daughter Maggie and family, owners of another frequented porch. Though architecturally dissimilar, the porches share many of the qualities that make any front porch a *grand* front porch.

Both have casual furniture and porch swings that beckon you to sit. Since both were built before building codes mandated thirty-six-inch-tall railings—a height that can make you feel like you're sitting in a playpen—both porches have railings low enough to see over and sturdy enough to sit on. Both have tongue-and-groove wood floors that slant to usher rain away and creak a little to remind you of the other visitors they've welcomed in the past. Both have ceilings made of bead board, a material formidable enough to stand up to the weather yet inexpensive enough to be used freely outdoors. At some point, one or both of these ceilings may have been painted light blue, a visual trick used to mimic the sky and lighten the interior of the house.

Both porches serve as a transitional space between outdoors and indoors—a place where you can drop your muddy shoes and muddled work problems before striding inside. Both are situated just the right social distance from the sidewalk to create that semi-public, semiprivate feel that helps define a successful porch. Most important, you'll find the occupants using their porches as a reprieve from the busy outside world and the more predictable world just inside the front door.

It's no coincidence that the decline of the front porch coincided with the rise of the automobile. Owning a car meant you were no longer tied to your community. You could work and shop twenty-five miles away, meaning you had less contact with the neighbor who formerly might have been your butcher or barber. You could drive to the beach, a park, or a friend's house to relax instead of rocking on the porch. You were enclosed in a four-wheel metal cocoon as you drove past porches with people you might have otherwise stopped to interact with if you'd been on foot.

Modern ranch houses, split levels, and "little boxes made of ticky tacky" eschewed by Pete Seeger had no time or space for front porches. People who had older houses with open porches often screened or glassed them in. House exteriors became more austere; garage doors the size of billboards replaced porches as the defining architectural element. Front porches were replaced by backyard decks and patios.

Telephones, radios, and televisions meant no longer having to rely on Madge down the street for the latest news. Air-conditioning eliminated the need to step (or sleep) outside to beat the heat. By the 1930s porches were gasping their last breaths.

Although the front porch lay comatose for decades, it wouldn't die. In the 1960s, Jane Jacobs, author of *The Death and Life of Great American Cities*, promoted the benefits of "eyes on the street," which included butts on the porch, which kept neighborhoods safer and helped build community. Thinking slowly shifted. In the 1990s, Robert Davis took eighty acres of Floridian "sand and scrub" and developed Seaside, a place the literature describes as "the world's first New Urbanist town."[3] It was a planned community consisting of shops, galleries, and three hundred homes—each with a front porch.

Other communities followed suit. Liberty on the Lake, an easy walk from our home, mandated every home have a front porch. It's nickname? Stillwater's New Small Town Neighborhood. Nearly two-thirds of all homes built today sport some type of front porch; in the South that number approaches 85 percent.[4] Millennials, more than any other age group, yearn for one. Though some maintain that many of today's porches are more an architectural element than a gathering place, the resurgence is very real.

"[The porch] helps contribute to a welcoming sense of community," Claude Stephens, founder of the tongue-in-cheek Professional Porch Sitters Union Local 1339, maintains. "It is the only place where you can feel like you are outside and inside at the same time; out with all of the neighbors and alone reading a book at the same time. The porch is a magical place where you are transported to a better state of mind and memories are born."[5]

What's more, as an investment, a well-designed front porch is the ultimate improvement to curb appeal; on the right house, and in the right neighborhood, it can yield an impressive payback. On average, a new porch—presuming it meets your community's setback codes—will cost about $20,000; that's a lot of lynx muffs and buffalo overshoes. But it's impossible to measure the social return on investment of a safer neighborhood, a chance to chat with neighbors, and a place that's a little bit yours and a little bit theirs.

ELECTRICITY

Birds on Wires
and Sparking Pliers

I WAS READY TO CUT AN OPENING FOR A NEW DOOR TO THE BED-room of one of my kids. I'd stripped away the drywall and clearly saw three electric cables crossing the path. I went downstairs and clicked off the breakers marked "KIDS, BEDROOM LIGHTS" and "KIDS, BEDROOM OUTLETS." To be safe, I clicked the "HALLWAY" breaker too. Back upstairs, I tested each outlet with a beeping electrical tester—only *sounds of silence*. I flicked all the light switches—only *darkness, my old friend*.

I cut through the first two cables without incident. Upon severing the third, however, I found myself flying backward, wire cutter sparking, hair on end like some Looney Tunes coyote. And *the vision that was planted in my brain, still remains*: "ELECTRIC BASEBOARD HEATERS!"

But here I am typing away, dumb and lucky in equal measure. For that one negative encounter with electricity, I can count millions of positive ones—but for the sake of this exploration, let's follow the path of my shocking experience from its place of origin, a sixty-million-year-old Wyoming coalfield, to a five-hundred-thousand-pound generator to a six-ounce seared wire cutter.

The electricity on our block is generated by the Allen S. King coal-powered plant—one of sixty-three-thousand power plants around the world.[1] When I stand on my roof, I can see the 785-foot chimney piercing the air.

"But what's inside?" I asked myself—and the Media Relations department at Xcel Energy.

In mid-April, I find myself standing on the roof of the King Plant, two hundred feet above terra firma, with plant director Brian Behm. He wants to start on the roof to provide the big picture. And it is BIG. To the west we see piles of coal so vast the earth scraper shoving it around looks like a Matchbox toy. To the east we see the mighty St. Croix River, which supplies water to cool the equipment. To the south we see the plant of Andersen Windows & Doors, one of the three largest window manufacturers in the world. And to the north we see my quaint hometown of Stillwater. In every direction we see wires. When I look up, I see, attached to the side of the chimney, a peregrine falcon nesting box. Over the past thirty years, along with nesting boxes at three other power plants, the King plant has been the maternity ward for 225 falcons.[2]

Behm, a blue-jeaned engineer comfortable in the company of both three-piece suits and greasy overalls, begins with Electricity 101. "In most cases, when you create electricity, you're taking thermal energy and turning it into mechanical energy, which produces electrical energy," he explains. Usually, the thermal energy is steam created by water that's been superheated by coal, natural gas, or nuclear fission. "Hydroelectric and wind power are different," Behm adds, "because they're taking the stored energy in water or the wind to create the mechanical energy, but it's the same idea." Solar is the exception.

The plant burns about two-and-a-half-million tons of coal per year—some twenty-five thousand railroad cars–worth—and can go through as much as three hundred tons per hour when operating full bore. Originally the plant burned high-sulfur coal from Kentucky and West Virginia brought in by barge. But when new environmental regulations kicked in in the 1980s, the plant switched to low-sulfur Powder River Basin coal, which is now hauled by train from Wyoming. Behm points to an area packed with railroad cars and explains how Union Pacific Railroad stations strings of forty cars on the side track. As the plant replenishes its coal stock, an Xcel locomotive pulls one of the strings through the "dumper building" where entire cars are turned upside down and emptied, one by one. The coal falls into a massive pit below. "Then we do that again and again and again," Behm says. Behm likes to keep a forty-day inventory of coal on hand "in case something happens." Today, he has enough coal for thirty-eight.

From there, a labyrinth of above- and belowground conveyor belts moves the coal to various places and piles. It's all eventually run through a building containing mammoth rotating hammers that pulverize it to the size of sea salt. The granular coal moves to hoppers in the upper level of the plant via a mile-long system of conveyor belts.

Behm and I also move inside, and as we wend our way down eleven sets of stairs past hoppers, gauges, and pipes, I'm again struck by the massiveness of it all. Nothing is small. Bolts are the size of beer bottles, wrenches the size of crutches. There are ten-thousand-horsepower pumps and water intake pipes so large one could drive a Chevy Suburban through one without dinging the mirrors. Even the workers seem big. When I ask Behm how many people work there, he quips, "Almost all of them."

Kidding aside, it strikes me as a place where neither the clumsy nor the impulsively suicidal should work. The end-of-life options are limitless. There are two-hundred-foot rooftops to topple from, steaming turbines to scald you, front-end loaders to crush you, cyclonic furnaces to incinerate you, electrical charges to vaporize you, mountains of coal to bury you—and enough noise so if you were to choose any of the aforementioned options, no one would hear you scream. But there are safeguards everywhere. "Sixty days ago, one worker got a sliver in his hand while pulling on his work boots," Behm says. Before that it had been years since the last accident. Given the perceived hazards, the workers' comp rates are no more than those of a high school janitor.

At one point, we cut through the control room, where two plant-equipment control-room operators sit in front of a dozen monitors. "These guys control everything from here," Behm says. They guide me to a pair of monitors that, in their words, "are display monitors only, so we can touch them without screwing anything up." Giving me a virtual tour, they point to pulsing diagrams of flue gas temperatures, water pressures, selective catalytic reduction doodads, and controls for a myriad of pollution-control devices. I ask them about the "Fish of the St. Croix" chart on the wall. They explain that the Clean Water Act dictates that the plant's cooling water intake pipes safeguard fish. "Even shad," one of them explains. "They're so fragile, they'll die if you just look at 'em. But our equipment screens 'em out, puts 'em back in the river, and they survive."

Until it underwent a $400 million upgrade in 2004, the King plant was deemed the third-worst polluter in Minnesota.[3] That's changed. Behm walks me through a separate basketball arena–size building where they operate the pollution-control equipment. "It's basically a chemical plant," he says. The most problematic emissions from burning coal—nitrogen oxide, sulfur dioxide, and mercury—are reduced to almost zero by flue gas scrubbers, catalytic reducers, and other pollution-control processes. Electrostatic precipitators remove 99 percent of the fly ash. "Anything you see coming out of that stack today is steam."

Behm sighs, "These old power plants are really something—but their days are numbered." When I ask him to elaborate, he explains that Xcel has goals of reducing carbon emissions by 80 percent by 2030 and being 100 percent carbon free by 2050. "It's what people want," he says. "You're going to see a lot more wind, solar, nuclear, and new technology—technology that hasn't even been invented yet." We have a long way to go. Today, 35 percent of electricity in the United States is generated by natural gas, 28 percent by coal, 20 percent by nuclear power, 7 percent by hydroelectric power, 6.5 percent by wind, and 1.5 percent by solar.[4]

When the 550-megawatt King plant is shuttered in 2028, as now planned, the 785-foot-tall smoke stack will be toppled. I— and the nesting peregrine falcons—will sort of miss the stately old landmark. I wonder what will take its place. Hopefully wind— quiet and graceful. Or not. The typical wind turbine generates 2 to 3 megawatts, meaning it will take 180 turbines to replace the output of the old coal plant. We could get by with fewer of the massive 5- to 10-megawatt turbines now in development, but those would *each* stand as tall as the old smoke stack. The thirty-three tons of copper each turbine requires will need to be mined, processed, and shipped from somewhere. The wind turbines will need to occupy anywhere from 800 to 30,000 acres for maximum efficiency, far exceeding the 150 acres the current plant sits on—but it's actually a moot point since the area has too many trees, hills, and people to be used as a wind farm.[5] The turbines—and the workers from the old King plant—will need to be located elsewhere.[6]

Solar would be safer and better, right? Yes, but according to calculations, it will require up to 10,000 acres to accommodate the solar panels needed to replace that missing 550 megawatts.[7] Some people will argue that those 10,000 acres—wherever they are— would be better used for biofuel production, for carbon dioxide-absorbing forests, or for producing food or housing for people in need. One researcher estimated it would take an area the size of Rhode Island and West Virginia combined to accommodate the solar panels needed to generate the megawatts consumed in the United States,[8] a proposition the citizens of Pawtucket and Parkersburg would hardly jump for joy over. Individual photovoltaic systems are a good option; the payback period now can be as low as four years. But is your roof oriented the right way? Does your area have enough sunny days? And do you have the twenty grand to make the initial investment?

We can all but nix hydroelectric power; it took forty-five years and an act of Congress—literally—just to get a bridge built one-quarter mile upstream. Nuclear? A grand total of one plant has been built in the United States since 1996. Natural gas? It's still a fossil fuel. Going carbon-free is a noble and necessary undertaking. But whatever replaces the coal plant won't be invisible, trouble-free, cheap, or without its own environmental impact. Just what *do* we say when stepping up to the mic at the utility commission's public hearing to voice our opinions about the future of energy?

The decision on how much power the King plant should generate comes from headquarters in Denver nine hundred miles away. Unlike most commodities—corn, coal, or natural gas—electricity can't be stored; it must be generated at the same rate it's consumed. Thus, power generation is coordinated across seventy-two power plants in eight states for efficiency; a grid that produces eighteen thousand megawatts and supplies enough power to service twenty million homes.[9]

Behm and I descend to the main floor and walk toward the turbines and generator, the workhorses of the electric generating process. "It's all about this," Behm says. "This is the money maker."

To power the money maker, the granular coal is fed into a cyclone furnace, where air jets swirl it, gas torches ignite it, and temperatures reach 2,500 degrees F. Water passes through the furnace in large pipes, then the boiling water feeds into a boiler tank, one so immense it's suspended from the ceiling so it can freely expand—as much as eighteen inches—at maximum temperature and capacity. The steam is compressed until it reaches four thousand pounds per square inch and then is blasted against the blades of three sets of turbines—papa bear, mama bear, and baby bear—each milking the steam for more and more power. As the steam cools and loses energy, it's condensed back into water and begins the journey anew. Water drawn from the river—never intermixing with the ultrapure turbine water—helps with the cooling and condensing. Once the river water has rested in outdoor cooling tanks long enough to bring the temperature down to 85 degrees F, it's released back into the river; it has only been borrowed.

"We keep the steam temperature at 1,000 degrees, plus or minus a degree, when it enters the turbines," Behm says. "Too much fluctuation throws the turbines out of balance and causes fatigue." And as he goes on to describe the intricacies involved in operating this colossal machinery, I realize we're standing inside an eleven-story, coal-powered Swiss watch.

A shaft from the whirling turbines extends to the generator, the sine qua non for the plant that engages two-and-a-half-million tons of coal, miles of conveyor belts, ninety employees, and enough water to float a battleship—and it's no larger than an eighteen-wheeler you see traveling along the interstate. Behm escorts me to the generator and tells me to put my hands on the housing. The rotor revolving inside at thirty-six hundred revolutions per minute contains coils of copper wire weighing half-a-million pounds—and the vibration is no more than our Vitamix at home, though the generator outputs 550 megawatts, or enough electricity to power half-a-million households and businesses. When I give Behm the "WOW that's a lot to absorb" head shake, he comments, "It all boils down to that steam coming out of a tea kettle driving a toy pinwheel experiment that you did in fourth grade."

People were aware of electricity long ago. Ancient "fish healers" would subject those suffering from headaches to a powerful jolt from an electric eel as a cure. And people noted that rubbing an amber rod with animal fur created some type of magical charge. In 1600, English scientist William Gilbert coined the term "electricus," derived from the Greek word for amber.

In 1752, Ben Franklin proved that lightning carried an electric charge with the celebrated "key on a kite with wet string" experiment—but the world wasn't in need of electric keys. It needed something that could replace the limited work capacity of animals, humans, and steam.

No one person invented or discovered electricity. Thomas Edison, Alexander Graham Bell, Nikola Tesla, George Westinghouse, and others built upon one another's discoveries to turn electricity from a mere quack curiosity into something that could be harnessed. Yet as far as we've come in the utilization of alternating current (AC) electricity, the mechanics of creating it—rotating coils of wire inside a magnetic field—have changed very little.

The battle between Edison's direct current (DC) and Tesla's AC—"the war of the currents"—is legendary. They were collaborators, adversaries, dueling geniuses, and eccentrics, each with varying degrees of business acumen (Edison won in that department). They competed to establish the first widely accepted electrical grid. Initially, DC took the lead, but it had its limitations: it was difficult

This display at the Smithsonian National Museum of American History highlights the history and versatility of electricity. The bulb—strikingly similar to the incandescent bulbs of today—was made in 1882 and employs a bamboo filament. The Gibson guitar was crafted 102 years later.

to transmit long distances. And separate generators and circuits were required for different applications—one set of wires for lights, others for motors and other equipment.

Alternating current was more versatile and could travel longer distances, but it was arguably more dangerous. To prove the dangers of AC, Edison initiated a misinformation campaign that went so far as to electrocute cats, dogs, and Topsy, a rogue zoo elephant who'd squashed three handlers in three months.[10] The 1893 World's Columbian Exposition in Chicago became the battleground, with the more versatile AC winning out.

Although Tesla won the technology battle, he eventually lost his marbles. Toward the end of his life he worked furiously on developing wireless AC and ray guns. He consumed only boiled milk and kept everyone three feet away to avoid germs. He was celibate, because, as he put it, "I do not think you can name many great inventions that have been made by married men." He wiggled his toes several hundred times before going to bed because he believed the nightly exercise would help him live to be 135 years old. He died at the age of 87, alone and nearly penniless.[11]

As we exit the King power plant, Behm explains that the plant consumes about 10 percent of the electricity it generates. He points out one group of massive power lines heading east across the river to Wisconsin; I follow another set west as I drive home. I have a new-found indebtedness to the wires, towers, and transformers I pass.

A typical high-voltage transmission wire is made of seven cables of steel wire (for strength) surrounded by eighty to one hundred aluminum cables arranged in a twisted pattern of concentric rings. These high-voltage transmission lines, strung along massive metal towers spaced every eight hundred to one thousand feet, are suspended from insulating disks of porcelain, ceramic, or other nonconductive material. The more disks an insulator has, the higher the voltage wire it's securing. If you see an insulator with four disks, the wire is most likely carrying 70 kilovolts; if it has fourteen disks, the wire is carrying 230 kilovolts. The largest insulators have sixty disks.

If you look closely, you'll gasp; the high-voltage lines are bare—not a lick of insulation. To shield the wires, the insulation would have to be so thick it would prevent heat from dissipating. Insulation would also increase the mass and weight of the power lines, making them more susceptible to wind- and weather-related damage; the increased weight would also mandate heftier towers spaced closer together. This nakedness is why they're strung so high.

There are three major power networks or grids in the United States: the Eastern Interconnection, the Western Interconnection, and the electric Reliability Council of Texas. The Canadian power grid also pokes its nose into the States.[12] One source calls this grid the "largest interconnected machine on earth," with two hundred thousand miles of high-voltage transmission lines and five million miles of local distribution lines.[13] The grids largely operate independently from one another, but in an emergency, power can be channeled from one to another. The Northeast blackout of November 9, 1965—started when a massive protective relay switch in Ontario tripped—affected thirty million people and required other grids to step up. If you were born in August 1966, you may be able to attribute your conception to an electrical malfunction near Niagara Falls on a lightless night nine months earlier.

As electricity nears its final destination, it branches out to substation step-down transformers. These are assemblages of voltage regulators, oil circuit breakers, switches, and lightning arrestors perched on metal frames, which are often surrounded by chain-link fences. These transformers lower the voltage more in line with the needs of surrounding communities and end users. Manufacturers and industrial complexes needing massive amounts of electricity may have distribution wires leading directly to them. Smaller wires, often grouped in threes, carry about 13,800 volts and travel over to your neighborhood atop poles usually made of wood.

Why wood? Though concrete, carbon fiber, and metal poles are available, nothing does the job cheaper, faster, or with more versatility than a stick in the mud. With a wooden pole, you have a sturdy support (average life span, forty-five years) which you can hang crossbars, power lines, cables, telephone wires, streetlights, traffic signs, and "Lost Schnauzer" posters—using little more than a drill and some hardware. That's why in this synthetic age there are 180 million wood poles standing duty in the United States.[14]

The lowest lines you see are usually low-voltage telephone and communication wires and cables. The four-foot open area above is a "safety zone" in which low-voltage workers can operate safely (the area below, affectionately called the "yard sale zone"). Next up the pole are electrical service lines and "service drops" (like the one that leads to your home). The very upper echelons are reserved for higher voltage distribution lines.

Some power lines—especially in large urban areas and near airports—are installed underground. Below street level, the wires are less susceptible to damage, take up less space, and are less affected by weather or fires—as PG&E's California wildfires will attest. The downside? Underground lines cost ten to fifteen times as much to install, have half the life expectancy, and can take weeks, versus days, to repair.[15]

Before branching off to your home, overhead power lines travel through a small cylindrical step-down transformer. In most cases, this transformer lowers the current to 120 or 240 volts. From there, the wires—at this point bearing insulation—enter your home through a service mast that directs them to your main service panel, which—if things go wrong—directs current to your wire cutter.

The average US household consumes 10,399 kilowatt-hours (kWh) of electricity per year, and the annual electric bill averages $1,500. Residents of Louisiana have the highest average annual consumption at 14,200 kWh, thanks to air-conditioning. Citizens of Hawaii, on the other hand, have the lowest at just over 6,000 kWh.[16]

Some people pooh-pooh household current, maintaining "it's not enough to kill ya," but they're wrong about two hundred times a year. Damaged or exposed household wiring, faulty power tools and appliances, ladders contacting power lines, and knucklehead moves like mine all contribute to the statistic.[17]

When I cut through the wire, my wire cutter created a "short" circuit for a split second, enough time for power from a "hot" wire to surge through my tool to a neutral wire. The worst route would have been for the current to travel from hand to hand, passing through my heart on the way.

Why, you might be wondering at this point, aren't birds on a wire electrocuted? Electricity follows the path of least resistance, and a big copper wire offers easier passage than a bird's skinny legs. Also, current flows only when a circuit is completed—and a bird on a single wire doesn't do that. BUT if a bird with a large wingspan touches two lines of different voltages at once, or a small bird touches a wire and a support pole that's grounded, the bird will indeed be zapped.

WALK THE WALK

Watts It Cost?

In the early days of electricity, some home systems were coin operated; you knew exactly what you were getting for your quarter. Not today. Here are some cost comparisons to ponder (amounts are based on electrical costs of $0.11 per kWh):

Refrigeration
1996 15-cubic-foot frost-free refrigerator: $198/year
Newer 17-cubic-foot Energy Star refrigerator: $46/year

Lighting
60-watt incandescent bulb: $90/10 years
7-watt LED bulb (comparable to 60-watt incandescent): $18/10 years

Cooling
Ceiling fan: 1¢/hour
Room air conditioner: 8¢/hour
Central air-conditioning: 32¢/hour

Birthmarks

If you're yearning for good reading material on a walk, check out the "birthmark" imprinted on the metal plate or branded into each utility pole. These birthmarks contain abbreviations and numbers that reveal the manufacturer: BPC (Big Pole Company); plant location and year of treatment: P 08 (Poleville, 2008); species: SP or RC (Southern Pine or Red Cedar); and class and size: 5-40 (class 5, 40 feet long).

This birthmark is applied at the factory and positioned so it can be easily read when the pole is buried the proper depth (which is 10 percent of its length plus two feet). This normally puts the birthmark three to five feet off the ground. The smaller—often oval—metal tags you see on the "road side" of many older poles are inspection tags bearing the year an inspection was performed.[18]

WATER

Towers, Faucets, *and* Meters

WHEN I WALK AROUND TOWN, THE ONLY EVIDENCE I SEE THAT THE water from my kitchen faucet arrives some way other than pure magic are the town water towers, the little covers marked WATER on the sidewalk, and the occasional fire hydrant. Beyond these random reminders, my town's water system—like yours—is largely out of sight, out of mind. Even my $21 monthly water bill is barely noticeable.

But like you, I drink, flush, wash with, shower in, and use on average about ninety gallons of water a day. Where does it come from?

To find out, I stroll the four blocks to the Stillwater City Water Department, a handsome brick building that looks the same today as it did when it was built one hundred years ago. When I ask manager Robert Benson whether he knows anything about our house's water history, rather than clicking on a computer, he strides over to a battered four-drawer oak filing cabinet and pulls out a faded six-by-twelve-inch manila envelope. He dumps an array of tattered papers on the counter, and as we sort through them, we find the work order that shows city water was run to the house on October 28, 1906, by licensed plumber Wm Pozzini. A hand-drawn sketch of the pipe's path was on the back. We also find shutoff orders, a waterline replacement sketch, and a plumbing repair bill for $9.38. It's hard to imagine a $9.38 plumbing bill, but even harder to imagine the thrill of having running water brought to one's fingertips in 1906, as much of a game changer as having the internet brought to one's fingertips a hundred years later.

Local history books show there was running water in town before 1906. In the 1870s, blacksmith C. H. Hathaway built a reservoir to catch the three hundred barrels of water per hour flowing from a spring on his property, which he then sold to adjoining businesses. His pump was operated "by means of a horse treadmill affair."[1] In 1880, Hathaway was put out of business by the privately held Stillwater Water Company, which pulled as many as one million gallons of water per day out of nearby McKusick Lake. One year, it pulled enough water to turn McKusick into a mere puddle. The untreated water ran through a system of wood distribution pipes, a wayward fish occasionally clogging the lines.

A flimflam man was responsible for the first deep water well, drilled in 1888. Noting green bubbles escaping from a pond, he imagined Stillwater sitting on an immense oilfield—and he on an equally large fortune. To lure investors, he captured the gas bubbles—most likely methane from decomposing plant matter—in a rubber bladder, then ignited the gas before their dollar-sign eyes. He raised enough capital to drill a deep well. He never hit petroleum, but he hit plenty of water, a fact the Stillwater Water Company did not overlook.

Over the next century, Stillwater's water system experienced the highs and lows of water systems everywhere. Early growth was driven by insurance companies' policies of offering lower rates to people who lived close to fire hydrants; some towns had hydrants on the street before faucets in the kitchen. The city of Stillwater purchased the system in 1911 to ensure residents had water pure enough for safe consumption and plentiful enough for fire prevention.

Today, my city's water system consists of a network of eight wells, storage reservoirs, and a half-million-gallon water tower with a brawny lumberjack balancing atop a log emblazoned on the side—all joined together by six- to sixteen-inch-diameter distribution lines. In an area blessed with deep, clean aquifers, treatment consists primarily of adding chlorine and a one-part-per-million dose of fluoride, then testing the water regularly.

To prevent freezing, all the distribution pipes and service lines leading to each house are buried at least seven feet deep. Our waterline froze because initially it was buried only five feet deep. Then we had compacted the bejesus out of the earth covering it— minimizing its insulation value—by driving over it while landscaping. Then we got a weeklong stretch of 10-below temperatures.

As they branch out into neighborhoods, the distribution pipes gradually decrease in size. The service line leading to our house is one inch in diameter and passes through a valve called a "curb stop," which allows the water department to shut off the water to our house in case of an emergency or unpaid water bill. (The curb stop valve is what's below the metal disks or square plates saying WATER near every house.) Turning the valve off requires a special long-shafted wrench, but it also requires a call to your water department since they're the only ones allowed to mess with it. As water enters your house, it passes through a meter, nearly all of which are read by a radio frequency device carried by a truck that drives through your neighborhood every few months. Not all citizens love the radio frequency system. When it was first installed in my town, one resident complained to the city council, "Every morning I wake up nauseous . . . I have resorted to sleeping in my kids' playhouse, which is outside and unheated."[2]

Not every place has such easy access to clean water. If you live in Riyadh, Saudi Arabia, your water is drawn from the Persian Gulf, desalinated, and then pipelined three hundred miles. If you live in Qaanaaq, Greenland, your water comes from icebergs towed to shore and melted in a smelter.[3] If you live in New York City, your water travels 125 miles from reservoirs in the Catskills, flowing through tunnels, over dams, and under the Hudson River in hundred-year-old aqueducts, and then passes through the world's largest ultraviolet disinfection facility on its way to your faucet.

To get a big-picture look at our world of water, I grab lunch with Stew Thornley, a health educator in the Drinking Water Protection section of the Minnesota Department of Health. "All the water in the world today is all the water that ever has been or ever will be," he tells me. "It just changes form and location." The water could be locked up in icebergs, oceans, aquifers, clouds, or a can of Coke. The water you brush your teeth with might have been muck in a dinosaur swamp one hundred million years ago. Water keeps getting recycled. How much treatment it requires depends on where in the cycle you tap into it.

Thornley explains that bacteria, viruses, nitrates, and harmful chemicals *have* to be removed, but creating 100 percent pure water is almost impossible. In fact, you don't *want* ultrapure water—unless you're in the business of manufacturing semiconductors. Ultrapure water is corrosive, tastes terrible, and if consumed, will actually leach minerals out of your body. Water, the "universal solvent," has a magnetic personality that naturally attracts minerals with gusto. What makes water "taste good" are the salts and minerals dissolved in it; many of them—like calcium, magnesium, potassium, sulfates, and iron—are critical for good health. Some people will argue that the best-tasting water mimics the composition of a water you consume by the minute—your own saliva.

Fire Hydrant Trivia

The term "fire plug" dates back to when most city water systems were made of wood. When a fire broke out, firefighters would dig down to the water pipe, drill a hole, and then allow water to fill the pit so they could pump or bail out of it. When done, they would ram a stick into the hole to plug it and leave the end extending above the ground so they could use the same hole for future fires. Thus, the birth of the fire plug.

Thornley tells me that the biggest development in public health in one hundred years was disinfecting water with chlorine in 1908. It was controversial at the time; chlorine was the chemical used to clean up morgues and operating rooms. But introducing it to water wound up knocking out waterborne diseases like typhoid, which saved millions of lives. Later, other healthy water initiatives were passed. Prompted in part by the Cuyahoga River catching fire in 1969—a river which, according to some, oozed rather than flowed—the Clean Water Act of 1972 was passed. It provided funding for cleanup and established regulations that barred businesses and municipalities from discharging pollutants, untreated sewage, and chemicals into surface waters and wetlands. The Safe Drinking Water Act of 1974 replaced a patchwork of state and local water regulations with uniform safe drinking water guidelines for the 150,000-some public water systems in the United States.

Thornley is a realist. He explains, for example, how Environmental Protection Agency (EPA) guidelines allow ten parts per billion (ppb) of arsenic in water since total removal is exorbitantly expensive. "People think we're putting a price on their health," Thornley says. "Five ppb would be better, but that would put half the water in the West over the limit. It's just not economically viable."

Thornley bemoans the plethora of chemicals entering our drinking water that defy filtering and treatment. Traces of pharmaceuticals, chemicals found in hand sanitizers, and PFAS (per- and polyfluoroalkyl substances)—dubbed "forever chemicals"—found in Scotchgard, Teflon, and firefighting foams are winding up in drinking water, where their presence has been linked to cancer. Studies have shown male sperm counts in the Western world have fallen by 40 percent or more, perhaps because of chemicals in plastics and estrogen compounds in birth control pills that have worked their way into water supplies.[4]

About 15 percent of us get our water from private wells, 25 percent from municipal wells, and the remaining 60 percent from rivers, lakes, and other surface waters.[5] If you look at a satellite image of the earth at night, you'll notice the highest concentration of light is, by far, along rivers and coastlines—bodies of water that provide transportation, sewage disposal, power generation, food, irrigation, and, of course, drinking water. Because treating surface water is considerably more complex than treating groundwater, I head to the McCarron Water Treatment Plant of the St. Paul Regional Water Service (SPRWS) to find out how it's done.

The process starts at an intake station perched on the shores of the Mississippi River twenty-five miles from the treatment plant. Two five-foot-diameter conduits carry water underground to a chain of lakes; it gushes into Charley Lake and then meanders through Pleasant, Sucker, and Vadnais Lakes. Pleasant Lake makes me want to scoop my hands and drink directly from it; Sucker makes me want to reach for the Cipro.

Minnesota—with its ten thousand–plus lakes, the mighty Mississippi, and semipredictable rainfall—has plenty of water, but in other parts of the world, not so much. In Mexico, three-quarters of the country relies on packaged water. In Uganda, 40 percent of the population travels thirty minutes or more to access clean water. The United Nations Environment Programme predicts that by 2030 half the world's population could be "water stressed."[6]

In the United States the most pressing problem is a crumbling infrastructure that, because of breaks and leaks, loses about two trillion gallons of treated drinking water annually.[7] Keeping ahead of the country's impending water crisis curve requires around $200 billion a year.

From Vadnais Lake, the water moves via conduits to the Mc-Carron plant; along the way ten wells can be activated to bolster the amount of water heading to the plant. It enters the plant underground, which is where I meet Jodi Wallin, who is in charge of public information. Wallin knows water. For the past seventeen years she has descended into reservoirs, ascended into water towers, immersed herself in water history, and ushered tens of thousands of people through the convoluted workings of the water plant. "Information is power," she says. "People make better decisions about their water usage when they know where it comes from."

Together, she and I wend our way through the 1920s-vintage building, past black and white photos of men with handlebar moustaches and scrunch-brimmed hats twisting Paul Bunyan–size wrenches. We enter a room where everything—pumps, hoses, storage barrels, and workers—is uniformly coated in a subtle yellow powder. This is where quicklime and alum are added to the incoming water; softening it and forcing suspended materials to collide and flock together to create larger particles called—appropriately—*floc*.

Via gravity, the water moves to the clarifier, a massive, twenty-foot-deep circular reservoir with a rotating arm that gathers the sunken floc. The floc is moved to the dewatering building, where it's consolidated into layers of solid material that's eventually hauled away to be used as soil amenders.

The water moves through chambers containing huge paddlewheels. To lower the pH level, carbon dioxide is added. The state-mandated fluoride is also added at this stage. Though adding a mere $1 per year to your water bill, the addition of fluoride helps with bone development in kids and reduces the number of cavities. But this step is not without its controversies.

During the "Red Scare" of the 1950s and 1960s, some people maintained it was a step toward socialized medicine; others claimed it was covertly sapping the brains of Americans, and the testicles of male American youth. More than a thousand Florida communities defeated referendums to add fluoride to their drinking water. Commie concerns aside, many European countries have discontinued fluoridating their water, with few negative effects. Picking up the slack are fluoridated toothpastes, mouthwashes, and salts.

HACKS & FACTS **Soaking Wet Hydrant Trivia**

You know those photos of city kids playing in "illegally" opened fire hydrants? There may be nothing criminal about it at all. In New York City, anyone eighteen years old or older can fill out a "hydrant spray cap" permit at their local firehouse. The cap reduces the hydrant flow from one thousand gallons per minute down to twenty-five, and—if certain conditions are met—someone from the firehouse will come out, install the cap, and turn the hydrant on and off for your neighborhood's temporary cooling pleasure.

From the paddlewheels, the water flows on to secondary settling tanks and then to filtering basins, where it seeps through three feet of granular activated carbon and four inches of sand, removing remaining sediment and molecules that might create unsavory taste and smell issues. The water then moves to a massive underground concrete reservoir, a sort of subterranean parking garage for H_2O. Chlorine and ammonia are added to further disinfect the water. Sodium hydroxide is added to minimize possible exposure to lead, which still exists in some pipes leading from the street into older homes. To ensure purity, the water has been tested a dozen times along the way.

From that underground reservoir, the water is pumped into a series of distribution pipes, which, if laid end to end, would create a 1,250-mile pipeline from St. Paul to San Antonio, Texas.

As we head back to the main office—with the water debacle in Flint, Michigan, still in the news—the conversation returns to lead pipe.

"Wanna see a chunk?" Wallin asks. We weave our way to the back of a storage room, past plastic Christmas trees and used office furniture. Wallin pulls out a plastic bag containing a foot-long gray pipe cut in two, lengthwise. She thinks twice before opening it, then sighs, "Oh what the hell; I can wash my hands afterward." She spreads the two pieces apart and points to the rust-colored coating on the inside. "This was the root of the problem in Flint," she says. Actually, the problem wasn't the rust-colored coating, but its removal.

She explains how in a properly maintained system, the sodium hydroxide added to the water coats the inside of the pipe with a protective barrier. When Flint began pulling its water from a new source and treating it in a plant that had lain dormant for thirty-six years, the chemical balance of the water was thrown out of whack, thanks in no small part to a series of miscues driven by finances

A section of lead water pipe with the protective coating that helps prevent lead from leaching into a home's water supply. In Flint, Michigan, the chemicals that help build this coating were mismanaged.

and mismanagement. Residents complained of foul-smelling water, skin rashes, and hair loss.

"Initially," Wallin says, "the water tested fine." But eventually, because the water was the wrong chemical composition, it dissolved the protective coating, exposing the drinking water directly to the lead pipe. The incidence of elevated blood lead levels in children tripled in some areas. One Flint pediatrician lamented, "Lead is one of the most damning things you can do to a child in their life-course trajectory."[8]

Wallin explains that none of the main distribution pipes in the St. Paul system contain lead, but about 10 percent of the service lines running to individual houses do. Lead pipe—flexible yet durable and easy to work with—was initially ideal for making such connections. SPRWS tries to replace four hundred to five hundred services per year, but even when that task is completed, lead can continue to be an issue since it's the homeowner's responsibility to replace any lead pipes located beyond the public right-of-way.

What's My Water Cost?

We use about twenty-seven billion gallons of water per day in the United States. Depending on where you live, you most likely pay between one-eighth of a cent to one cent per gallon. The figures below are based on a cost of one-half cent per gallon. Here's where it goes and what a typical "use" might cost.

WATER USE	GALLONS PER USE	COST PER USE
Toilet		
Low flush	1½ gal/flush	<1¢/flush
High flush	6 gal/flush	3¢/flush
Running toilet	200–2,000 gal/day	$30–$300/month
Shower/Bath		
Low-flow showerhead	2 gal/min	5¢/5-min shower
High-flow showerhead	5 gal/min	12¢/5-min shower
Typical bath	40 gal/bath	20¢/bath
Dishes		
Newer Energy Star dishwasher	6 gal/load	3¢/load
Older-model dishwasher	16 gal/load	8¢/load
Hand wash dishes	8–27 gal/load	4¢–13¢/load
Faucet with slow drip	10–25 gal/day	$18–$45/year
Washing Machine		
Newer washers	25 gal/load	12¢/load
Older washers	40 gal/load	20¢/load

Source: "Water Q&A: How Much Water Do I Use at Home Each Day?," USGS, https://water.usgs.gov/edu/qa-home-percapita.html.

"Some houses were partially plumbed with lead pipe," Wallin says. "The solder used to join copper pipe had lead in it up to about 1986. Many faucets manufactured up until 1997 had lead in them too." And there are weird scenarios. For instance, she tells me about one customer who complained about high lead levels in his water. When water officials made a home visit, they found that the homeowner—in an effort to save a few dollars—had used lead-based solder to join dozens and dozens of short scrap pieces of copper to make one long pipe.

But back in Stillwater, my $21 a month buys me clean, plentiful, easy-to-use water, a necessity and luxury taken for granted.

We never know the worth of water until the well is dry.[9]

HACKS
& FACTS
Even MORE Fire Hydrant Trivia

Since many early water systems and fire departments were privately owned, the fire plugs were closely guarded. The companies would hire the meanest, ugliest, baddest-ass person in the neighborhood to serve as a guard. He was "plug ugly."

Today, the handle for turning on a fire hydrant is the five-sided nut on top of the hydrant. The odd number of sides requires a special wrench, which makes it harder for kids who want to cool off on hot days to mess with them.

4

MAIL

First-Class Diamonds, Babies, *and* Rattlesnakes

I'M QUITE CERTAIN I HAVE A HUNDRED FRIENDS BECAUSE EARLIER this year I dropped a hundred "Happy New Year!" cards in the big blue box in front of my post office. The marvel isn't that I have a hundred friends. Or that not one was RETURNED TO SENDER. Or even that the card sent to Chuckie in California arrived the same day as the one I sent to Erik and Kathy three blocks away. The marvel is that I've been sending letters, checks, and packages in the mail for fifty years and haven't the slightest clue how they get from point A to point B.

This is why I'm standing with an industrial engineer with the United States Postal Service (USPS) in the Minneapolis Postal Service Processing and Distribution Center (P&DC). The facility handled my one hundred cards, plus another one million other pieces of mail on an average day, heading into, and out of, the Minneapolis area.

It's midmorning, and the four-story building has only a moderate buzz. But at 4 p.m., when mail starts coming in, and 4 a.m., when mail is being shipped out, and in between—when every letter, magazine, envelope, and box will be sniffed, sorted, scanned, coded, canceled, re-sorted, bundled, and re-re-sorted—it will be a madhouse of computerized efficiency.

My guide is good at making the complex sound simple and has the polished crispness of a McIntosh apple. He begins by explaining that each of the 159 million delivery points in the United States is assigned an address, each address is assigned a ZIP code, and every ZIP code is assigned to one of the 191 P&DCs operating across the country in 2019,[1]

The Minneapolis P&DC handles all the mail leaving from, or entering into, ZIP codes that begin with 553, 554, and 555. Every evening, mail being shipped *from* ZIP codes beginning with those three numbers is dropped off at the P&DC. Mail with ZIP codes starting with those same three numbers—mail that's been sorted at the other 190 P&DCs around the country—is also shipped *to* the Minneapolis facility for final sorting and delivery.

The primary mission of this P&DC is to sort all the mail destined for its three ZIP codes so it's easy for local carriers to deliver it—and to sort everything else into 190 other clumps to be flown or driven to the other P&DCs around the country. It's akin to gathering a half billion jumbled marbles of 191 different colors, sorting them into 191 boxes, and then getting those marbles into the right pockets of 159 million kids—each day.

At the expansive loading dock, semis and vans unload hampers of mail starting in the late afternoon. The hampers, containing a jumble of letters, manila envelopes, and boxes, are dumped onto a conveyor belt where the contents are manually and mechanically "culled" into three categories: standard envelopes, flats, and boxes. This is where I realize that labeling a package THIS SIDE UP is a fool's game.

For expedited processing, Priority and Express mail are shuttled onto separate tracks. The machine involved in the culling process is affectionately known as Barney, because of its similarity in color to the purple cartoon dinosaur. "Weirdly enough, Barney sorts out one or more wallets a day," my guide says. "People find one, don't know what to do with it, so they drop it in a mailbox."

Lots of other strange things get dropped in the US mail. Ripley's Believe It or Not! holds an annual contest to determine the weirdest unpackaged thing that can be shipped to its headquarters. In a recent year a traffic cone, prosthetic arm, rural mailbox, and tree trunk with embedded horseshoe all arrived . . . unpackaged (but with the correct amount of postage). It's perfectly legal.

Things are also removed from mailboxes. "Mail fishing"—lowering glue-coated bottles or sticky mouse-trapping paper into mailboxes, then extracting envelopes to steal cash, gift cards, or checks—became so endemic in New York City that the USPS began retrofitting or replacing all seven thousand of its curbside mailboxes with ones that have narrow slots. One fishing expedition on 83rd Street had netted $53,000. Actually, many curbside mailboxes are being removed altogether. In 2000 there were 400,000 across the country; by 2015, there were fewer than 160,000, with the number going down every year.[2]

As we meander toward another colossal machine—a convoluted thing, part wheat combine, part Cray supercomputer—my guide pauses and cranes his neck upward. "What do you think is in those big square ducts?"

"Air?" I venture.

"Nope. Postal inspectors. Maybe. We never know how many are up there—one, a dozen, none . . ."

The dinky windows I spot reinforce his words. It turns out most post offices of any size have a "lookout gallery" with its own dedicated entryway.

"How about the small round ducts?" he continues.

I consider replying, "Short postal inspectors."

"Anthrax sniffers."

After the anthrax attacks of 2001—which killed five, including two postal workers—this step is required if the USPS wants to ship mail on commercial airlines.

My guide continues to explain that the postal inspector's mandate is to "protect the sanctity and security of the mail." They take their jobs seriously. Fourteen postal inspectors have been killed in the line of duty. They're part of *the* oldest law enforcement agency in the US and can carry firearms, serve search warrants, and make arrests. (Postal inspectors accounted for nearly six thousand arrests in 2018.)[3] Their authority includes prosecution of cases involving mail and internet fraud, cybercrime, Ponzi schemes, identity theft, narcotics, mail fishing, counterfeit postage, and child pornography. They also tracked down the Unabomber in 1996.

We arrive at the convoluted machine that sorts the first-class letters. The Advanced Facer-Canceler System uses optical character recognition (OCR) technology to scan each envelope, and a quick-draw mechanical hand orients the address face forward and right side up. Each envelope slaloms to another computer, which reads the address; compares it to a database to make sure the recipient, address, and ZIP code jibe; then prints a yellow fluorescent bar code on the back. The machine verifies that the stamp is a stamp, then cancels it. The time it took for this machine to process all one hundred of my Happy New Year cards clocked in at sixteen seconds. The artificial intelligence embedded in the OCR correctly reads more than 95 percent of the addresses zinging past it. The easiest to read are typed envelopes; the hardest are letters to Santa scrawled in crayon.

When the computer detects something is amiss, it scans the envelope and sends the image to the Remote Encoding Center in Salt Lake City, where a human being reads and manually keys in the address. If this agrees with the database, the letter zips back into the system. If not, the envelope heads to be further processed manually; the worst cases head for the national Dead Letter Office.

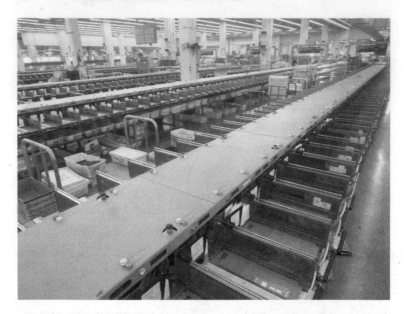

This Delivery Bar Code Sorter reads and collates mail at a rate of thirty thousand pieces per hour and arranges letters by delivery sequence for your carrier—with pinpoint accuracy.

The mail zings onward. Another OCR reads the fluorescent bar code on the back and prints a black bar code on the front. Each envelope then travels to another machine, which sorts mail into "destinating" bins, which stay within the 553, 554, or 555 ZIP codes—like my card to Erik and Kathy—or "originating" bins destined, like my card to Chuckie, for one of the other 190 P&DCs.

I ask my guide about the mind-boggling speed and accuracy of these 1990s putty-colored machines. "They *are* 1990s machines," he says. "They're beasts. They're durable. They get the job done."

The next machine takes "originating" mail and sorts each envelope into one of 190 bins—one for each P&DC around the country. By early morning, each of those 190 bins will be shipped to the designated P&DC.

The "destinating" mail—as well as the incoming 553, 554, and 555 mail from the other P&DCs—runs through a Delivery Bar Code Sorter (DBCS) three times. The first time, the machine sorts it into batches based on the five numbers of the ZIP code and shuffles them into trays. Each tray holds about four hundred envelopes. This is called sorting to the ZIP-code level.

The next two passes further refine the sorting process so that by the end the letters have been sorted by address and the order the carrier will deliver them. Even the card sent to Erik and Kathy had to run this sixty-mile gauntlet—complete with sifting, sorting, and stamping—to reach its destination three blocks away.

In a different part of the building, the flats and packages follow the same approximate process. Small boxes that can be processed by machine are separated from large boxes that are handled almost entirely by hand. Magazines are shuffled into appropriate bins in another part of the facility.

Like first-class mail, flats and boxes are sorted to the ZIP-code level, but unlike mail, not every piece is presorted to more refined levels. In such cases, the final sorting is done by carriers at the destination post office or carrier annex. Many businesses, publishers, and direct-mail companies presort their mail to the three- or five-digit ZIP-code level before dropping it off. The further one sorts, the cheaper the postage.

The same technology that's greatly increased the sorting rate has also greatly decreased the employment rate. "A generation ago, there were well over two thousand people working at the facility," my guide says. "Today there are about nine hundred—six hundred in mail processing, two hundred in maintenance, and one hundred in transportation." This human sparsity is a condition that I note not only at the P&DC, but at water-treatment, sewage-treatment, and power-plant facilities as well. Rather than scores of steel-toed laborers pulling levers, toting carts, and lifting bales, you're more likely to find a handful of Skechers-footed employees glued to banks of monitors or hooking up diagnostic computers. I pause to wonder whether we've simply exchanged one form of monotony for another.

As I consider this, trucks are loaded with letters, flats, and packages, and off they go.

It's 5 a.m.

I ask Elizabeth, our postal carrier, whether I can tag along someday. She thinks this is an okay idea. I shadow her one steamy summer afternoon. When I ask her about the "neither snow nor rain . . ." adage, she tells me that in her eighteen years as a carrier she can remember only two times when *they* were "stayed from their appointed rounds." "Once was last year when the temperature hit 30 below—and that was on my day off. The other time was ten years ago when it snowed so hard the roof of the Metrodome [the football stadium in Minneapolis] collapsed." Elizabeth works on a rotating schedule—never on Sundays, off Monday one week, Tuesday the next week, and so on—meaning, she gets two days off in a row less than once a month. On her days off, a "utility person"— usually someone paying their dues until a route opens up—fills in.

Elizabeth starts her day at 7 a.m. at the local postal carrier annex along with forty other carriers. She corrals her presorted letter trays and then grabs her catalogs, magazines, and packages and arranges them so they too can be delivered in sequence. Hers is a "park and loop" route, where she drives to an area, delivers mail by foot in a loop, then drives to the next area. Those with "mounted" loops spend their day in a vehicle delivering rural and city mail.

Elizabeth drives a classic, boxy Long Life Vehicle (LLV). Before 1918 most carriers walked their routes. When World War I came to a close, the USPS inherited a fleet of army surplus vehicles consisting of forty-five different makes and models—a mechanic's nightmare. By the mid-1940s the fleet was in shambles. In 1957, the USPS introduced the Mailster—a three-wheeled, two-cylinder, 7.5-horsepower fiasco so lightweight it could bog down in three inches of snow or be knocked over by a large dog. In the late 1950s, Jeeps, which had shown their meddle during World War II, were pressed into service. Switching their steering wheels to the right side alone allowed drivers to shave sixty to ninety minutes off their route times.

At the beginning of each walking loop, Elizabeth grabs a fistful of letters; an armload of magazines, catalogs, and flyers; dutifully locks her LLV; and then heads down the block. There are door slots, domed rural mailboxes, and quaint old-fashioned flip-top boxes. Occasionally she "shoots" a package with her GPS-equipped Mobile Delivery Device (MDD) before dropping it off. The MDD doesn't track just packages but Elizabeth as well, leaving digital breadcrumbs supervisors can use to track her progress. (One supervisor, in a different state, discovered that a carrier was having an affair with the supervisor's wife when the MDD showed the carrier making regular half-hour visits to his home.) The device can also issue alerts for road conditions, dangerous animals, missing children, and civil unrest.[4]

One house on her route is being remodeled, and the front stairs have been removed. Elizabeth wends her way through the debris and delivers the mail into the box, six feet off the ground, with an astute overhead dunk shot. She greets Lilly the cat with a tummy scratch, quizzes Talus an Alaskan malamute as to the whereabouts of her yard mate, and greets everything on two or four legs with a pat or hello. Even squirrels' tails twitch when they see her coming. Some postal workers, though against policy, distribute dog treats along the way. "It's better to be a friend than a stranger," Elizabeth explains.

Since being bitten by a German Shepherd, she keeps a can of animal repellent on her hip. She's not alone; of the thirteen thousand mail carrier injuries incurred in a typical year, six thousand are from dog bites. Houston, Los Angeles, and Cleveland have the highest incidence of dog attacks.[5] Jose Salazar, a Houston carrier, has been bitten nine times. Another carrier had an index finger severed by a dog while stuffing mail into a door slot. Other hazards—slips, falls, vehicle accidents, repetitive motion injuries, and physical attacks—abound. The USPS ranks fourth in companies with the largest number of serious accidents, just behind the blade-wielding chicken processors at Tyson Foods.[6]

Wherever she goes, she's greeted with polite honks, waves, picture window nods, and chit-chat. During her eighteen years on the route, she's seen residents come and go, toddlers turn into teens, heads of hair turn from brown to bare, and addresses once bearing both "Mr." and "Mrs." now bearing just one name.

At the end of her route, she has a small handful of odds and ends. "Every once in a while, the sorting machine has a bad day, or there's someone new running the machine—then I'll have a pile of undeliverable mail that needs to be fed back into the system. But most days it's okay."

What does Elizabeth do after walking seven miles delivering mail? "I take the dog on a three-mile walk. I love walking; but you'd have to pay me to run."

Over the years the USPS has delivered mail via stagecoach, roller skate, motorcycle, skis, Segway, toboggan, dogsled, biplane, supersonic jet, snowmobile, and hovercraft. Pneumatic tubes once delivered six million letters a day, including half of New York City's mail. Mules are still used to deliver mail the eight miles to the bottom of the Grand Canyon.

The package Harry Winston used to ship the Hope Diamond—with its alleged curse—to Washington, D.C. In the year following the delivery of the diamond, the carrier's wife died, his dog strangled on his leash, his house was destroyed by fire, and he was involved in two car accidents. *Inset:* The Hope Diamond.

It's hard to beat the reliability of the USPS. Harry Winston had such faith in the system that when he donated the $350 million Hope Diamond to the Smithsonian in 1958, he shipped it by registered mail—in a plain brown envelope stamped FRAGILE—from New York City to Washington, D.C. The cost was $145.29 for postage (which was only $2.44 of that amount) and $1 million in insurance.[7]

In 1913 Mr. and Mrs. Beagles of Glen Este, Ohio, mailed something even more precious—their eleven-pound grandson—to a relative who lived a mile away. Postage was 15 cents. To prove they were doting grandparents, they insured him for $50. In June 1920 the *Los Angeles Times* carried this tidbit: "Children may not be transported as parcel post, First Assistant Postmaster General Koons ruled yesterday. . . . Mr. Koons said children clearly did not come within the classification of harmless live animals which do not require food or water while in transit."[8]

The USPS uses 600,000 workers, 31,324 post offices, and a fleet of 232,372 vehicles to deliver 150 billion pieces of mail a year. That's 47 percent of the world's mail.[9] At 55 cents per letter (in early 2020), the US has neither the cheapest nor most expensive postal service. To mail one of my Happy New Year cards in Norway would have cost $1.78, in Bangladesh less than a dime.[10] Post offices can be found at the base of Mount Everest, on the Antarctica peninsula, and nine feet underwater in the Republic of Vanuatu, where snorkeling tourists can mail special waterproof postcards.

In 1913 an eleven-pound child was mailed to relatives for 15 cents. The service was discontinued when it was determined children did not meet the standard of "harmless live animals which do not require food or water."

The streamlined system for delivering my Happy New Year cards was developed over two centuries. In How the Post Office Created America, author Winifred Gallagher posits, "The history of its post office is nothing less than the story of America."[11] In the beginning, the states of the United States were anything but united. The Puritans of Massachusetts, the Dutch of New York, and the elitist plantation owners of the South had

HACKS & FACTS

Stop Junk Mail

If you're an average Joe or Joan, you receive forty-one pounds of junk mail a year, don't open 44 percent of it, and recycle about a quarter of it. There are a variety of services you can use to stem the flood. The Data & Marketing Association offers a service that allows you to opt out of some direct marketing for ten years for a $2 processing fee (https://dmachoice .thedma.org). The Paperkarma app allows you to snap photos of a piece of unwanted junk mail and then tap SEND and UNSUBSCRIBE to opt out of future mailings. An internet search using the words "stop junk mail" will provide more services for ending direct mailings and other junk mail.

---

banded together to throw off the shackles of British rule, but with that done, they found themselves a collection of disparate colonies. Add to this a flood of pioneers heading west and a system of roads that consisted of little more than rutted trails and you had a country that was very loosely stitched. What was the needle that could sew them all together? The Post Office.

The mission of the newly formed postal service was to encourage an open forum for the exchange of ideas, provide an affordable way to send newspapers and books to promote literacy, and cast a lifeline to pioneers heading into the wilderness. By 1809, the United States had twenty-three hundred post offices, which quickly became gathering places and hubs of the community. There were hiccups; people who were enslaved were prohibited from working in the postal system lest, while working, they run across documents proclaiming "all men are created equal."[12]

Instituted in the 1860s, home delivery required city names to be formalized, streets to be named, houses to be numbered. An address created a tangible identity that many, up until then, had lacked.

The Railway Mail Service, where mail was sorted while trains were en route cross country, dramatically shortened delivery times. Railway clerks were expected to have photographic memories, to heft two-hundred-pound mailbags, and to be able to sort mail into pigeonholes for hours while standing in a rocking train car.

Different levels of service—first, second, and third class—were instituted. A certain Charles Lindbergh was one of the first Air Mail Service pilots. Zone Improvement Plan (ZIP) codes were established during the 1960s. The first of the five numbers indicates an area, and as you go east to west, that number gets larger. The second two numbers indicate a region within that area, and the final two numbers identify a specific post office. The "+4" numbers narrow the delivery point even further. In 2000, 103 billion pieces of highly profitable, highly mechanized first-class mail were delivered.[13] What could go wrong?

Well, the internet could come along. One consequence of the 2001 anthrax attacks was that people who had previously been wary of email converted to using that technology. The internet also changed the way people paid bills, banked, shopped, made reservations, researched, and sent Happy New Year greetings. The personal, emotionally charged act of handwriting a letter was replaced by the slapdash posting of an emoji. The volume of first-class mail—once the largest profit-maker for the USPS—started dropping at a rate of 3 percent a year, cumulatively resulting in a 47 percent decrease in volume by 2020. Today, this drop in revenue has been partially offset by the 10 percent annual increase in the shipping and packages business—the "Amazon effect." Still the service loses around $3 billion a year.[14]

Today, the USPS is an independent agency regulated by the US government, but it receives exactly $0 in funding—a sort of Twilight Zone for any business. On the package delivery front, the USPS is being challenged by FedEx, UPS, Amazon, drone deliveries, and self-driving delivery trucks. While mail volume is decreasing, the number of addresses is increasing by about four thousand delivery points a day. And, while the USPS does have a legal monopoly on delivering standard and first-class mail, it's hampered by its mandate to deliver mail at a minimum level (five or six days per week) at a uniform price. This is profitable in large cities but a money-losing proposition in Hooper Bay, Alaska, where mail must be flown in by bush plane.

Yet, in an era when we've become accustomed to instant electronic communication, receiving an honest-to-god letter, thank-you note, or Happy New Year card is a simple pleasure that will persist—through rain, sleet, and snow.

Dead Letters and Live Snakes

As the postal service grew, so did the need for dealing with letters and packages that were undeliverable and nonreturnable. In 1825, the Postal Service opened a Dead Letter Office in Washington, D.C., to deal with the occasional "lost and found" item. By the turn of the century, more than twenty thousand items a day entered the center.[15]

By 1850, so many bizarre items were discovered that an informal museum was created. One newspaper reporter compared it to "the handiwork of the inmates of a lunatic asylum."[16] Items included mummified fruitcakes, dentures (which one visitor recognized and claimed), a lock of hair from Charles Guiteau (the man who assassinated President Garfield), and the preserved bodies of three rattlesnakes.

The story goes that a perforated can containing the three live rattlesnakes arrived at the Dead Letter Office. The superintendent requested that the Smithsonian send "someone accustomed to handling such reptiles" to chloroform them. Upon completion of the task, the dead snakes were left in the open can. Later, a female visitor to the facility spotted a "rattlesnake coiled and ready to spring at her." Fortuitously, a letter carrier threw his mailbag onto

As long as there's been mail, there've been dead letters and packages. This photo, taken around 1922, reveals one corner of the Washington, D.C., Dead Letter Office.

the snake, then beat the living daylights out of it with a fireplace poker, saving the damsel in distress.[17]

Also legendary in the Dead Letter Office was Patti Lyle Collins, who could decipher and decode illegible addresses with 99 percent accuracy. She was said to know the location of every street, college, and corporation in the United States and most other countries. Those in Collins's position would need to deal with letters like one from Mark Twain, addressed to "Mr. C. M. Underhill, who is in the coal business in one of those streets there, and is very respectably connected, both by marriage & general descent, and is a tall man & old but without any gray hair & used to be handsome. Buffalo N.Y. P.S. A little bald on the top of his head."[18]

Over time, a dozen dead letter offices were opened across the country, but in 2011 they were consolidated into a single Mail Recovery Center (MRC) in Atlanta. The people who work there are part detective, part appraiser, and part auctioneer. The center receives ninety million items a year. Items that are deemed to be valued at $25 or more (about twelve million of them) are examined to see whether a recipient or sender can be located. About 20 percent of the time MRC detectives are successful. Items that can't be traced are retained for 30 to 180 days and then recycled, destroyed, donated to nonprofits, or auctioned. (If you've lost an item, you can begin the search process by visiting https://www.usps.com /help/claims.htm.)

One painting was found to have $5,000 worth of marijuana hidden in the frame. But most don't have that kind of value and are clumped into themes and then auctioned by the boxful. A recent lot contained everything one might need for a quiet evening at home: a My Pillow, a Monopoly game, a combination hairbrush-hairdryer, and a "Satisfyer Pro 2 vibrator."[19]

TELEPHONE WIRES AND WAVES

From Tin Can to iPhone (to the Bieber)

AS WE WALK AROUND THE BLOCK, WE SEE SIGNS OF COMMUNICA-tion equipment everywhere: overhead wires, towering antennas, crouching junction boxes, and manhole covers with conduit snaking below. These are the devices that keep us instantly apprised of important news events—like reports of pop star Justin Bieber peeing into a mop bucket in a New York City restaurant one raucous night.[1]

It took only a millisecond for news of the "mop bucket incident" to travel the 1,020 miles from New York to my cell phone. But how long would it have taken this information to travel given past modes of communication?

Foot Messenger: Eleven Days, Twenty Hours

In 1986, Stu Mittleman set this record for an endurance run of 1,000 miles, while averaging three hours of sleep per night.[2] Philippides ran 26.2 miles from the battlefield of Marathon to Athens to announce victory. Upon arrival, he collapsed and died, highlighting one of the drawbacks to this form of rapid communication.

Horseback: Three Days, Twenty-One Hours

Carrying a copy of President Lincoln's inaugural address, the Pony Express covered the nineteen-hundred-mile route from St. Joseph, Missouri, to San Francisco in seven days, seventeen hours. The process involved switching horses every ten to fifteen miles and switching couriers every seventy-five to one hundred miles.[3]

Signal Telegraphy: One Day, Twelve Hours

Early signal telegraphy used fire, drumbeats, flags, reflected light, and other signals to transmit simple information along a series of stations or towers. Around 800 CE, during the Tang Dynasty in China, a message could be sent along the Great Wall at a rate of seven hundred miles in twenty-four hours.[4]

Messenger Pigeon: One Day, Ten Hours

"Wayne Jr." established this record for a one-thousand-mile race in 1927—a record that still holds today. Wayne Jr.'s average speed was 1,122 yards per minute.[5]

Optical Telegraph: Four Hours, Ten Minutes

This method employed shutters or flags to transmit coded letters, words, and messages from point to point. In 1811, the system was used to send information of the birth of Napoleon's son from Paris to Strasbourg, a distance of 247 miles, in one hour.

Telegraph: Three Minutes

A skilled telegraph operator could send about sixteen words per minute in 1850. Thus, a fifty-word missive describing the mop bucket incident would have taken about three minutes.

Switchboard-Based Telephone: Forty Seconds

In 1930, an experienced operator could gather the call information, fill in the record ticket, look up the route number, and plug in the appropriate cables to connect a long-distance call in less than a minute.

Cell Phone: 0.0053 Seconds

Most digital and wireless information travels at slightly less than the speed of light, which makes the transmission of most information instantaneous. Calls between those using the same network travel slightly faster than calls traveling on different networks.

I wasn't around for the Pony Express era; but I was around for the tin-can era, the dial-phone era, and the cell-phone era. Listen in.

For those who had play-deprived childhoods, the tin-can phone is made by punching holes in the bottoms of two tin cans, inserting the ends of a string into each hole, and tying knots. When the string is pulled taut, one can becomes a primitive microphone vibrating up to one thousand times per second, and the other becomes a primitive speaker.

In the mid-1800s more than three hundred patents were filed around this acoustical telephone concept; transmissions of up to half a mile were possible. There were, of course, problems. The line had to be held taut to transmit the vibrations, making turns and long distances problematic.[6] Also, one needed to have a separate line for each communicant; taken to the extreme, this method would turn the skies into a schizophrenic spider web of string.

What was needed was a system that could transmit vibrations long distances, along with a central switching station that would connect callers. Telegraphs that worked by transmitting electrical impulses had been around for a while; if there was a way to send vocalizations instead of dashes and dots, you'd be getting somewhere. Perhaps you could call this device a "teletrefono."

The challenge was how to combine the acoustical benefits of a tin can with the electrical improvements of a telegraph. In the 1850s, a German schoolteacher created a "mechanical human ear" that could transmit and receive the nuances of the human voice. He stretched a sausage skin across an opening to imitate the tympanum (or "drum") part of the ear, then positioned a thin curved lever of platinum wire (the "hammer") against the sausage skin, so when the skin vibrated, this lever rapidly opened and closed an electrical circuit.

Antonio Meucci developed an electromagnetic device he called the *telettrofono* in 1871. The sketches and a working model he sent to the Western Union lab where Alexander Graham Bell worked inexplicably disappeared shortly after their arrival. Through a series of blunders Meucci lost the claims to his patents and devices. Bell subsequently registered *his* patent for an electromagnetic telephone that bore many similarities to Meucci's. William Orton, president of the Western Union Telegraph Company, once said he found it amusing that Bell is viewed as the man who spent his fortune defending his telephone patent while in truth he spent his fortune patenting the work of others.[7]

While the "transmitting" problem was being solved, the process of connecting multiple callers was being tackled. All the telephone wires in an area were routed to a building called a "telephone exchange," which housed operators. When Mildred Brown wanted to talk with Henrietta Smith, the operator would physically connect the two lines with a short patch cord to "make the connection." And if the operator wanted to listen in through her headphone, she could and often did, making her the epicenter of local gossip.

Switchboard operators working at an early telephone exchange. Emma Nutt—the first female operator, hired in 1878—reportedly knew every phone number in the New England Telephone Company directory by heart.

US NATIONAL ARCHIVES @ FLICKR COMMONS

Eventually human operators were replaced by automated switching devices. Early dial phones would emit nine clicks when the number nine was dialed, four clicks when the four was dialed, and so forth. Like a cylindrical combination lock that opens only when the right numbers are lined up, the connection between phones would "open up" only when all the right sequence of numbers had been dialed. Eventually, rotary phone "clicks" were replaced by push button "beeps" of varying tones.

In 1880, fifty thousand telephones were in use. Would they catch on? It looked doubtful. A committee from Western Union determined: "We do not see that this device will be ever capable of sending recognizable speech over a distance of several miles. . . . The idea is idiotic on the face of it. Furthermore, why would any person want to use this ungainly and impractical device when he can send a messenger to the telegraph office and have a clear written message sent to any large city in the United States?"[8]

Oops.

By 1910 there were close to six million phones in the United States. Brick and mortar telephone exchanges were replaced by smaller automated exchanges. By 1970, more than 90 percent of households in the country had a landline phone. But as people and technology got busier, the idea of a phone with no strings attached became alluring.

That could change things a bit.

Work on the first wireless phone took place in the '70s—the 1870s. Alexander Graham Bell had imagined a "photophone" that would transmit sound via light beams;[9] his concept failed, but the notion persisted.

Cell phones are essentially high-tech walkie-talkies. Like tin cans and landlines, your cell phone contains a microphone and a speaker. The microphone is essentially a teeny "broadcast booth" for sending voice waves; the speaker is sort of a teeny radio for receiving radio waves. It's these "old-fashioned" radio waves that allow modern cell phones to operate.

So, let's say you're a Belieber and you want to personally ask him about the mop bucket incident. You dial 1-800-BIE-BER1. Your brain sends an electrical signal to your vocal cords, which send a sound wave to a microphone, which turns that wave into an electrical signal, which turns that into a digital signal that's transmitted via radio waves to a tower with a receiver, which converts that digital signal back into an electrical signal that's turned back into a sound wave that goes through Justin's complex inner ear system, which turns that into an electrical impulse that goes to his brain. As your call pings from tower to tower to tower, Justin's phone is sending out its own signal looking for yours. And then they meet. And you begin your conversation, "Justin, about your choice of urinals . . ."

HACKS & FACTS

Cell Phone Phobias

Nomophobia (no-mobile-phone phobia) is the fear of leaving home without one's mobile phone, of being out of cell phone range, or of having one's battery die.

This is a vast—VAST—oversimplification. For instance, what is the "cell" in cell phone? The term refers to the patchwork of small areas or cells that land is divided into, each containing its own tower or base station. The cells are blob-shaped, though they're often depicted as hexagonal. In rural or suburban areas, each cell can be ten square miles or larger, but in urban areas—where many more calls are being made—the cells are smaller. Specific bands of radio frequencies are set aside for cell phone use. "Set aside" isn't exactly correct since the Federal Communications Commission (FCC) auctions off spectrums of frequencies; in a recent year, this brought in $45 billion from companies like AT&T, T-Mobile, and Dish. Any time you use your cell phone, it's working on two frequencies—one for sending and one for receiving. Cell phone transmission isn't all wireless; fiber-optic cable, Bluetooth wireless technology, and good old copper wire often play roles in the process.

Cell phone signals are purposefully weak—about half a watt. A "weak" signal means it's transmitted to only the nearest cell tower. The need to send out only a wimpy signal means your cell phone can have dinky batteries and be the size of a Kit Kat bar rather than a toaster; the first mobile handset weighed 4.4 pounds.[10]

In the United States in 2018, there were 325,000 "cell sites" and base stations—two-hundred-foot cell phone towers and backpack-size base stations mounted on utility poles.[11] Towers can cost up to $250,000 a crack to build, so whenever feasible, companies lease space on tall buildings, water towers, billboards, smokestacks, even church steeples. One tower in my neighborhood is disguised as a pine tree—with scurvy. One superbly located urban site leases for $150,000 annually, but $10,000 to $20,000 is closer to the norm.

Some experts estimate that there are four hundred million "connected devices" in the United States and eight billion cell phone subscriptions in the world—more than the number of people in each respective sphere. Most developing countries have just skipped the whole landline thing. In a remote area of Tanzania where I've worked, 2 percent of households have landlines while more than 75 percent of adults have cell phones. In an area where there are few banks, libraries, newscasts, or doctors, cell phones have become all these things.

Given the speed, volume, and accuracy of the cell phone network—coupled with the fact that it's used to transmit movies, text messages, internet data, and everything else—it's quite mind boggling, really.

On the other hand, it's surprising how simple, in spirit, it's all remained. When Samuel Morse sent a telegraph message in the 1850s, he sent a series of electronic dots and dashes; when we use a cell phone today, we're sending out digital dots and dashes.

HACKS & FACTS

The Colorful Tale of "Bluetooth" Gormsson

King Harald "Bluetooth" Gormsson was known for *uniting* the warring tribes of Denmark and Norway in 945 CE—thus, the origin of the term *Bluetooth*, the technology *uniting* wireless phones and other devices today.

PART II

OUTGOING

RECYCLING

From A+ to D in One Hour Flat

EVERY OTHER MONDAY MORNING I WHEEL A SIXTY-FOUR-GALLON jumbled mess of Coke cans, wine bottles, milk jugs, shredded paper, and refried bean cans to the curb—then righteously rub my hands together knowing I've done my part to save the planet. An A+ performance. I all but wait for Mother Nature to applaud. I know the contents of my bin will magically be reborn as cardboard boxes, maintenance-free park benches, rebar to build new cities, Phoenix from the ashes. But after trailing my bin across town to a recycling facility, I learned that (1) there's nothing magical about it, and (2) my performance is, at best, a D. I may even be an accidental arsonist.

I'm standing with Bill Keegan, president of Dem-Con, a waste and recycling management company, and Jennifer Potter, community outreach coordinator, in a classroom attached to the Materials Recovery Facility, or MRF (rhymes with "Smurf"), as they enthusiastically tag team facts and stories of the recycling world—and building—we're about to enter. "In this industry," Keegan tells me, "success used to be measured by no one knowing you were there. If there were no complaints about noise, smell, or traffic, that was *all* you wanted; *that* was success. But there's been a paradigm shift. Now we *want* people to know we're here. We're proud of what we do. We want to bring people in to show them the story."

Keegan tells the tale of one city council meeting where people were up in arms about the prospects of a landfill and recycling facility going in to their community. "One woman was screaming at us. Telling us how we were destroying the planet and giving her kids asthma attacks," Keegan explains. "I politely asked her where her waste went, and she responded, 'To the curb.' Well, we want to show people where stuff goes when it leaves the curb."

In its short six-year lifetime, the massive facility has undergone seven major upgrades. Such is the chameleonic nature of recycling. Dem-Con welcomes thousands of visitors every year, targeting fifth-graders who are young enough to be impressionable, old enough to understand, and unjaded enough to spread the gospel. The company has an interactive trailer they haul to schools and festivals along with 3D reality glasses that allow kids and adults to virtually step into the recycling process. A sorting station invites kids to practice hands-on recycling. The goals are to help build a culture that makes recycling natural and automatic at a young age and to trigger a feeling of unease in people when they *don't* recycle something.

I'm no fifth-grader, but I'm surely a gold star student when it comes to recycling. Keegan, Potter, and I don our blaze yellow vests, hardhats, and safety glasses and head into the facility, where I intend to confirm my stellar grade.

We climb thirty steps to a platform overlooking the tipping floor. Collection trucks back up and, in four- or five-minute intervals, disgorge mound after mound of mixed recyclables. The space is the size of an ice arena with a corresponding warmth; all is rumbling, screeching, hand signals, and shattering glass. Keegan shouts over the growl of trucks and machinery: "People recycle twice as much stuff with commingled or single-stream recycling like this [where all recyclables are collected in a single container] than when people sort their own stuff into bins." He continues, "It also builds good habits. People roll their bins down to the street and realize the garbage bin is only half full and the recycling is full, and they start feeling okay about their trash."

Dem-Con processes more than fifty thousand pounds of commingled recyclables per hour; of that, 8 to 10 percent is garbage or items that should have been put in "the other" bin. "People are well-meaning," Potter says. "They have something they're unsure of and figure if it goes into the trash, there's zero chance it will be recycled, but if it goes into the recycling bin, well, maybe. . . . It's called 'wish-cycling.'" But those wishful items wind up contaminating other materials, jamming machinery, adding to the amount of hand sorting, and can be downright deadly. The best mantra, according to Keegan, is "When in doubt, throw it out."

Every few minutes a steel-wheeled front-end loader takes a crunching bite out of the twenty-five-foot-high mountain of recycling and dumps it into an infeed hopper. The hopper shakes the material onto rollers that spread it evenly onto a wide conveyor belt. Keegan explains that bagged recycling needs to be ripped open by hand. When handlers don't have the time to do that, the entire caboodle is diverted into the trash stream. Plus, the bags clog machinery and aren't recyclable at that plant. Right off the bat, my grade drops a notch.

Plastic bags, large and small, not only clog Dem-Con's conveyor belts, they're clogging the planet. Five trillion plastic bags are produced worldwide each year; only 1 percent are recycled.[1] The average "working life" of a single-use plastic bag is twelve minutes;[2] tests show it can take a hundred years or longer for one to decompose.[3]

It's easy to get lost in the cornstalk maze of recycling statistics. For every argument, there's a counterargument. Paper or plastic? Well, it requires about a thimbleful of petroleum to create one plastic bag. BUT getting a paper bag from forest to checkout counter requires four times that amount of oil. AND since cotton farming is so pesticide and irrigation intensive, you need to reuse a cloth bag 173 times for it to "pay off."[4]

Keegan, Potter, and I follow the four-foot-wide conveyor belt as it travels uphill to a sorting cabin where six eagle-eyed, quick-fisted workers perform recycling triage. They pluck out oversize, dangerous, or blatantly unrecyclable items. It's dizzying.

The sorting process is 80 percent mechanical and 20 percent human, but literally every item in the three hundred tons of recyclables that pass through the plant daily has human eyes put on it.

I peek into the rejects tubs and find lamp bases, twisted copper pipes, shattered light bulbs, and old faucets. "You can't believe the stuff we find in here," Keegan shouts above the clang and roar. "We find tens, twenties, hundreds. If we find a wallet, we return it. Hundreds of dirty diapers. If there was an inventor's competition for the worst thing to be recycled, it would be the diaper—a plastic thing, with elastic bands and cotton holding crap—not one part recyclable. And they don't smell that great on a hot day."

Potter continues: "We've pulled out guns, safes, deer parts. One time we found a turtle crawling on the belt." (They rescued it.) In other facilities workers have found lawnmowers and hand grenades. "Sharps, syringes, and medical wastes are a constant problem," Potter adds. "Everyone along the line has a pull cord and can stop the line—and it stops the whole line—to safely remove things."

As Potter speaks, one worker extracts a bowling trophy with one hand and a bra with the other.

The team at Dem-Con is eternally vigilant, looking out for lithium-ion batteries, which are used in everything from vape pens

A front-end loader consolidates some of the three hundred tons of commingled recyclables dumped at the recycling plant each day. About 8 to 10 percent of the stuff entering the plant should have been thrown in the trash bin.

to cell phones to cordless drills. Last year, one of Dem-Con's transfer stations burned to the ground because of a single crushed lithium battery. It's not unusual for a dozen fires a year to break out in the plant we're touring. The batteries also cause most collection truck fires.

Recycling and refuse disposal is the fifth most dangerous occupation in the United States. Most fatalities occur during the collection process, but plenty occur in the plant. One Florida worker was crushed by a cardboard compactor. Another worker in Albany, New York, was killed while removing a piece of plastic from a conveyor belt. Workers are injured by falling bales of materials, are caught in shredders, and contract hepatitis from loose needles.

One worker scrunches his forehead and yanks out a string of Christmas tree lights and a busted dinner plate—both items I've wish-cycled. I'm suddenly not feeling so grandiose about my recycling choices. My grade slides further.

As we move down the line, Keegan and Potter hoot out first names, joking and jiving with workers. Dem-Con is a third-generation family business, started in the 1960s by "Grampa Pahl." "Back then, landfills wanted *more* stuff," Keegan says, "but Grampa Pahl's descendants saw the future. They realized recycling, processing, and landfill diversion were more sustainable models." In 2010, the company had 17 employees; by the end of the decade, that number had grown to 180.

As we move on, a bag of shredded paper gets caught in a roller and explodes like a confetti cannon. Thousands of shreds rain down on people, equipment, and recyclables. "That's my number two evil," Keegan sighs. "Look at it. This looks like a parade route. It's impossible to clean up. It can contaminate an entire batch of glass or plastic. It jams equipment."

My office shredder and I plead guilty, and my grade drops another notch.

A massive conveyor belt moves to a series of spinning disks that launch cardboard panels and flattened boxes into a massive bin; everything else falls through. At the highest level, the plant is designed to separate two-dimensional materials (primarily cardboard and paper) from three-dimensional items (primarily containers). Because balled-up paper behaves like a 3D container, it makes it more difficult to sort. Likewise, a "steamrolled" milk carton behaves like a 2D object, making that more difficult to sort. Potter says it's best to keep containers at least roughly in their original form. I realize that the masses of balled-up wrapping paper we toss into our recycling bin at Christmas not only impede the process, but the stuff with glitter, a metallic coating, or velvety flocking isn't recyclable to begin with.

The containers go through a machine that smashes the glass bottles. Those shards are sent to a company that removes contaminants and then uses an optical scanner to sort the pieces by color. The shards are reborn as bottles, abrasives, or reflective beads for road paints and tapes. Everything else continues down the line. "The only glass people should recycle are those that originally had food or beverages in them," Potter explains. "Mirrors, window glass, and bakeware have a different composition and melting temperature and contaminate the truly recyclable glass."

A ton of recyclable glass costs glass-bottle remanufacturers less than the price of a pizza. But recycling one ton of glass saves ten gallons of oil, a seemingly trivial amount until you multiply it by the fifteen thousand tons of glass Dem-Con recycles yearly.[5]

Glass gets recycled in other ways. In 1960, Freddy Heineken, after encountering both homelessness and a plethora of discarded Heineken beer bottles on a visit to the Caribbean island of Curaçao, developed a square beer bottle with interlocking edges that could be used as a brick. Several prototype houses were built, but the idea never took off.

We enter an area where conveyor belts crisscross like an LA freeway. The materials pass over another series of disks that pull loose paper up and out of the stream and let everything else— mostly cans and plastic containers at this point—roll downhill to the container line. Keegan explains how a few years ago mixed paper and junk mail bales sold for $70 a ton and could contain 2 percent contaminants. Then the primary importer of plastics and paper, China, upped its standard to 0.5 percent contaminants and eventually stopped importing recyclable materials altogether, choosing to focus instead on recycling its own refuse. The US market tanked, but not before Dem-Con had invested $2 million in machinery to attain the higher standards. Now Keegan pays $5 a ton to get rid of it. "The spreadsheet from year to year can look kind of funny," he says. "Recycling has no environmental or economic benefit if you don't have an end market for it."

Recycling is a market-driven industry, and when demand drops or costs become too high, it can take a step back fifty years. In 2019, Philadelphia was hit with sky-high price increases for its recycling program and, rather than absorb or pass along the costs, chose to incinerate half of its recycling-bound materials. Similarly, "dirty recycling," coupled with other contamination issues, led to every ounce of recycled material deposited at the Memphis International Airport winding up in a landfill.[6]

We continue zigzagging along a series of ramps and walkways. You don't simply see the process; you smell, feel, hear, and taste it. We pause at the only sorting machine that, to me, makes intuitive sense—a magnetic drum that pulls my refried bean and tuna fish "tin cans" off the line and hurls them over an imaginary goalpost into another bunker.

We follow the main line to a six-foot-long optical scanner that blasts light onto each plastic item zinging by. The information is fed to a computer that in milliseconds determines each container's composition. The data are sent to a series of air jets that blast No. 1 plastic bottles—a fraction of the one million per minute the world generates—into the appropriate bin.

We enter an enclosed room where Potter introduces me to Hot Dawg, an artificial intelligence robotic sorter right out of a sci-fi movie. When installed, it was only the second of its kind in the industry. It consists of a frenetic tentacle that snatches and sorts out milk cartons, juice boxes, and No. 2 translucent plastic containers with the accuracy of a cartoon lizard tongue zapping flies out of midair.

We move farther into the room where a dozen workers stand hip to hip before a relentless conveyor belt, further hand sorting No. 2 and No. 5 plastic containers. Containers with stuff stuck to the inside are tossed into the trash line. Potter explains how spending ten seconds scraping peanut butter from the inside of a jar at home helps on multiple fronts. "The optical scanners can 'read' and sort the container more easily, food doesn't contaminate the pure plastics, and clean containers mean eliminating the temptation for homeowners to store recyclables in a plastic bag at home; it can make the difference between a jar being resurrected as a lawn chair or spending its life in a landfill."

On we march, past the aluminum can sorter. Though the cans aren't magnetic, an electrical current hurdles them through the air. Recycling royalty, aluminum cans are in high demand, because they can be infinitely recycled and bring in $1,400 a ton. Recycling a ton—or seventy-five thousand cans—can save enough electricity to power an average home for a decade.

We follow a branch line carrying loose paper and junk mail. When I ask Keegan about the conspicuous lack of newspapers, he holds up his smartphone and explains, "We've had an 85 percent reduction in newspaper since these things took over."

At the far end of the building, we look down to see massive compacting machines creating Mini Cooper–size bales of material. "People think their recycling has value and that we should be paying them for it," Keegan explains. "But it's a mixed bag. Aluminum is absolutely valuable, but it's offset by the paper and glass that bring little or may even have negative value." He surmises that the market value of a ton of mixed recycling—once sorted—might be $35 today—a steep dip from the $140 it brought ten years ago. But when you subtract from that $35 the processing costs—the hauling, the sorting, the shipping—the price per ton can look pretty dismal. One solution to the economic conundrum is instituting extended producer responsibility (EPR) laws where companies that produce or use plastic and other containers pitch in on the cost of recycling programs, as they've done in Canada and Europe.

We're getting better. In 1960, Americans recycled 10 percent of their waste; today, we recycle 35 percent.[7] My home state of Minnesota, along with other states and cities, set an ambitious target to compost or recycle 75 percent of waste by 2030; San Francisco is already there.

We climb into Keegan's truck to tour the rest of the campus. We visit the construction debris area, where dumpsters of wood, shingles, bricks, and metal are sorted. After unusable items such as insulation and drywall are picked out, about 75 percent of the materials find a new home—some of it as granular cover that's daily layered onto landfills for sanitary purposes. We pass a machine that looks like a Tyrannosaurus Rex chewing up two-by-fours and pallets and spitting out wood chips that will become landscape mulch, animal bedding, and erosion-control materials. "I have about a fifty-fifty chance of getting out of here without a flat tire," Keegan grins. Glancing to the right I see a shivering, flannel-shirted, semi driver crouched before a punctured front tire whose chances are zero.

Shingles are ground into rice-size particles and stockpiled. Some of the particles will be integrated into asphalt mix for roadwork. Dem-Con has worked with the Department of Transportation to develop a "road topping" gravel-asphalt shingle blend that reduces dust and rutting on rural roads by 80 percent. For another end-market use, Dem-Con pulverizes wood waste into biofuel for powering the nearby Rahr plant, the largest beer malting operation in the world. "The more beer you drink," Keegan quips, "the more you're supporting recycling."

We pass the vehicle recycling facility, where dismembered cars sit amid piles of brake drums, rotors, tires, and barrels of fluids; all will be recycled. Potter points to the stacks of wrecked cars needing to be stripped. "You get a bad winter, you get a lot of crashes. We got a lot of cars last week."

Keegan and I pop our heads in to visit Dan Chilefone, general manager of the metal recycling division, and find him walking between massive cardboard bins filled with copper, wire, and pipe of every shape and size. When individuals bring in metal—and plenty do—they snap photos of the person, their ID, and their load. "There's a lot of theft in this industry," Keegan explains. "If someone reports a big spool of copper cable has been stolen, and Spike, you're standing there with a big spool of copper cable in your truck bed, well, you've got problems."

I ask Chilefone about the most precious metal the company recovers, and he leads us to a pile of vehicle catalytic converters, each containing $50 to $200 worth of rhodium, platinum, and palladium. When I ask about the weirdest thing brought in, he and Keegan turn toward each other and chant, "Hawaii chair." We traipse into a back room where an odd-looking office chair sits plugged into the wall. They have me sit in it as they turn it on. As I twirl around like a gawky hula dancer, another worker walks in laughing—and I realize that the company, from plastic plucker to president, has that just-right blend of work and play. There's a family feel to it. It's 10 below zero outside, but the room feels warm.

Sweden—which arguably recycles 99 percent of its trash—is considered by many to be the paragon of recycling. The "arguably" comes into play since Sweden counts incinerated garbage (energy recovery) as part of the equation. Indeed, Sweden imports more than two million tons of trash annually to keep its incinerators burning and power plants running. We can learn a lot from the Swedes; citizens are meticulous in sorting their own recyclables and depositing the materials at stations rarely located more than a quarter mile apart. Composting is encouraged. Ads are run on television. It's become part of the Swedish mindset; they feel BAD if they don't recycle. Skol!

Recycling is a mixed bag in the Americas. In the United States, each person generates 4.5 pounds of solid waste per day and recycles about a third of it. Canada recycles about 30 percent, while Mexico recycles about 5 percent. Worldwide, Austria comes in No. 1 with a 63 percent recycling rate; Japan, an island nation, has a surprisingly low 21 percent rate.[8] Alabama recycles about 9 percent of its trash, one of the lowest rates in the nation.[9] With an 80 percent recycling rate, San Franciscans are in the vanguard.[10] Most of us sit halfway between Alabama and San Francisco with our recycling habits—and the beauty of recycling is it's a matter of choice for each person, household, and community to select which direction to drive.

I drive home to my recycling bin, report card tattered and smudged. Teacher's comments include: Does not follow directions (though my recycler sends me a list of dos and don'ts every year). Does sloppy work (it wouldn't take much time to rinse that thing out). Needs to show more effort (I could easily compost that shredded paper).

But I'm gonna bring my grade up.

**WALK
THE WALK**

Five Worst Things
to Throw in the Recycling Bin

Bill Keegan's been in the recycling business for twenty-four years and has seen it all. Here are five things he'd just as soon never see again (but sees hourly):

1. *Lithium-ion batteries* can explode and catch fire when broken or crushed by machinery. They're responsible for fires nearly every month at Dem-Con. SOLUTION: Visit https://www.call2recycle .org and punch in your ZIP code to find a battery drop-off center near you.

Bill Keegan of Dem-Con, a recycling and waste management company, holds an example of a lithium-ion battery responsible for starting the fire that destroyed one of the company's transfer stations.

2. *Shredded paper* can contaminate entire batches of glass and other materials; it's slippery underfoot, is a cleanup nightmare, and can't be recycled at Dem-Con (though it can at some centers). SOLUTION: Compost, minimize shredding, and take paper to local shredding events.

3. *Plastic bags, films, and tarps* wrap around belts and rollers, stopping equipment and entire production lines. SOLUTION: Deposit bags in drop-off boxes located at the entrances of most chain stores. Visit https://www.plasticfilmrecycling.org for exact locations.

4. *Hoses, wire, and extension cords* tangle machinery and can be a danger for workers to remove. SOLUTION: Repair broken hoses and extension cords; otherwise, trash them.

5. *Sharps and syringes* pose health hazards for recycling workers; conveyor belts must be stopped to remove items safely. SOLUTION: Visit https://www.safeneedledisposal .org to learn about your state's guidelines for disposal.

Note: Diapers are No. 10 on Keegan's list—but No. 1 on most line workers' lists. Procter & Gamble and other companies are working on diaper recycling—but for now the only place for the 450 billion diapers used each year worldwide is in the trash.

SEWERS

The Lifesaving World
of Wastewater Below

AS YOU SIDESTEP THE MANHOLE COVERS ON YOUR WALK, YOU'RE probably not ruminating about your first job interview—unless it was with Microsoft in its infancy. According to company legend, interviewers asked candidates, "Why are manhole covers round?" The purpose? To loosen up the candidate and check his or her ability to think outside the box, so to speak. We'll get to those "whys" soon enough, but for now let's look at the story of what flows below—a story best told starting not at the beginning but at the end.

I've undergone security clearance and shown my government-issued ID. I'm wearing closed-toe shoes and long pants. I've donned my visitor's badge and protective eyewear. I'm ready for today's mission impossible: making it through a tour of the Metropolitan Wastewater Treatment Plant in St. Paul without having an out-of-stomach experience.

I've asked Matt Simcik, professor at the School of Public Health at the University of Minnesota, to give me a primer as to what to expect. "All sewage treatment systems—whether it's a mammoth Chicago treatment plant or a rural septic system—work in the same fashion: they take the process that would normally happen in nature, then accelerate it. And there are lots of ways of accelerating it." In even more simplistic terms, it involves taking out most of the stuff in the wastewater that isn't H_2O—especially anything that will chew up oxygen in the river, lake, pond, or ocean it's released into.

The plant—the tenth largest in the United States—is perched in a hardcore industrial area along the banks of the Mississippi River. Our guide for the afternoon, Raymond Smith, ushers us into a room and begins by showing us a diagram of the facility—a convoluted chart that looks like an abstract Paul Klee mural. But as the day goes on, things become less abstract.

Smith begins at the beginning. When pioneers began settling the area, they disposed of their sewage in the then-conventional manner—in outhouses, fields, and nearby bodies of water—which was fine since the waste from forty-four hundred people in 1850 amounted to little more than the proverbial drop in the bucket. But as the population grew, so did the need for a more systematic and healthy way of disposing of sewage. In 1880, the first sewers were built, discharging their contents directly into the Mississippi River.

In 1926, the U.S. Bureau of Fisheries conducted a water quality survey of the Mississippi and found only three living fish in the forty-mile stretch downstream from St. Paul.[1] Sewage had killed the river. Water guidelines were established, and by 1938, the facility that we will soon be strolling through was in operation. Today, the plant treats 250 million gallons of wastewater generated by nearly two million people daily. The plant employs three hundred people and is so vast that street signs mark the intersections of service roads.

We enter the plant at the same place as the *influent*, the wastewater destined to be treated. The massive building features a series of three-story-tall welded steel combs with ten vertical teeth spaced half an inch apart. Like the influent, the first two stories are below-ground. This bar screen's mission is to strain out any large solid objects heading into the plant, specifically objects that could damage the pumps and other equipment. (Feces, toilet paper, and garbage disposal discharge have largely become liquefied at this point.) "We see a lot of strange things," Smith says. "Rubber duckies, cell phones, underwear, stacks of bills stapled together, Happy Meal toys—lots of Happy Meal toys." One worker quips, "I won't go in for anything under a hundred." Some people at other plants have removed bowling balls and live pythons. Every few minutes, scrapers lick the bars clean and deposit the goop onto a conveyor belt, which delivers the debris to a dumpster destined for a landfill.

Everything at the plant moves by way of good old-fashioned gravity. The influent flows to settling tanks or grit chambers where—spurred on by aerated water—sand, smaller solids, and seeds settle to the bottom; these dregs are scraped, dried, and then also hauled to a landfill. Smith pauses our tour to explain that over the past twenty-five years the amount of flow through the plant has decreased, driven primarily by the separation of sewage and storm sewers and water conservation programs. (The city of Detroit is now spending $500 million separating sewage and storm sewer systems.)

The water moves into primary treatment, which consists of a series of massive aerated pools, where the fat, grease, scum, and renegade plastics rise to the top and are mechanically skimmed off. Remaining solids sink to the bottom, and a Ferris wheel of slow-moving cross bars pushes the sludge into one of two directions. Some is shuffled off to the solids management building. "We're not going into the solids building," Smith declares. "You don't *want* to go in there." Other sludge returns to the treatment process as *return activated sludge*. In an ironic twist, this sludge is reintroduced into the wastewater process to "eat" pathogens, bacteria, and organic matter that could use up oxygen in the Mississippi River, the wastewater's ultimate destination. Sated, the microorganisms "floc" together and settle to the bottom, which lets the plant remove them. The entire operation is a delicate chemical balance that must be carefully monitored and maintained. Smith talks about protozoa, amoebas, rotifers, and absorption—things I was supposed to be learning in seventh-grade biology but was too busy trying to get Patty Burns to go to homecoming with me. But now I realize the purpose of these microorganisms—they eat shit.

Halfway through the tour, we stop at the water quality laboratory, a pristine palace perched alongside a river of excrement. The facility takes in more than seventy thousand water samples per year and conducts two hundred thousand analyses. White-coated technicians test samples of incoming influent for chemicals the plant may not be able to process and outgoing plant *effluent* for purity. They test wastewater from eight hundred industries that dispose of their waste at the plant, along with septic tank waste delivered via "honey" pump trucks each day.

With a knowing wink, our lab guide announces, "Each sample is given a social security number—which stays with it from the cradle to the grave." We make stops at seven different testing stations. There are spectrometers, refractometers, and -ometers of every sort. I'm touring with eleven people from the Minnesota Department of Health—many of them lab rats—and they begin peppering the chemists with questions. Enough acronyms to choke a lexicographer are thrown around: PVC, BOD, VOCs, PCBs. By the time we leave, I have only a vague notion of what's been discussed—but I feel supremely safe knowing my effluent is in good hands and well-labeled test tubes.

The effluent continues to flow on to secondary treatment, which consists of Olympic pool–size aeration tanks where oxygen and the aforementioned return activated sludge are introduced. Aerobic and anaerobic conditions are created, nitrogen is given and taken, biophosphorus removal takes place. There's more biological and chemical treatment, final settling, and clarifiers. Pathogens don't have a prayer—but other things do. Chemicals from certain pharmaceuticals, herbicides, antibacterial soaps, and other products don't succumb to conventional treatment; there's growing concern as these compounds wend their way into drinking water sources—the long-term impact unknown.

The final sludge from the process—about 3 percent solids—is thickened, centrifuged, dried, and then burned to power the plant in summer or heat the plant in winter. Sludge from treatment plants in other states is processed into fertilizer pellets. Sludge from New Jersey, the Garden State, is hauled 1,450 miles by truck and train to Tyler, Texas, and as recently as thirty years ago, New York City dumped its sludge—six million tons of it annually—106 miles out into the Atlantic Ocean. Up until the late 1980s, Los Angeles dumped its sludge into Santa Monica Bay via a seven-mile-long pipeline.

After its twelve-hour journey through the facility, the water is released into the *outfall* area, a channel that flows into the Mississippi River. The facility runs 24/7/365, and there's no OFF switch. During the warm summer months, when people swim or play in the Mississippi, the water is given an extra dose of chemical purification. From October to April, when only the criminally insane enter the river, extra treatment is withheld. All of this—from start to finish—for about 66 cents per day per household in the Twin Cities.[2]

When I asked Smith whether the outfall water is clean enough to drink, he answers, "Would I drink it? No. But it's safe enough for human contact." That's good. Fourteen million people down-river rely on the Mississippi as their source of drinking water.

Not all plants release their treated wastewater into rivers. One relatively new attempt to kill two turds with one stone is the "toilet-to-tap" approach—treated wastewater is reintroduced into ground-water reservoirs, most often ones that are rapidly depleting. Before being pumped back into the ground, wastewater goes through an intense purification process that includes reverse osmosis filtering and exposure to ultraviolet light. For ten years, Orange County in California has used its "Groundwater Replenishment System" to supplement the water supply for 850,000 residents.

Public perception is a tall hurdle. A recent blind taste test con-ducted by researchers at the University of California, Riverside, asked people to sample three types of water: bottled, tap, and toilet-to-tap water. Those participants with general "openness to experi-ence attitudes" ranked the three samples equally. Those more anx-ious about new experiences (in the designated "neuroticism" camp) actually preferred the toilet-to-tap water over regular tap water.

The Metropolitan Plant exemplifies state-of-the-art sewage treatment. But how did the process evolve? Deuteronomy 23:13 provides one early guide: "And you shall have a trowel with your tools, and when you sit down outside, you shall dig a hole with it and turn back and cover up your excrement" (English Standard Version). But for a more in-depth look, a trip to Paris is in order.

If the wait at the Louvre to see the *Mona Lisa* is too long, make the twenty-minute walk to the Paris sewer museum, Musée des égouts de Paris, located near the Alma Bridge crossing the Seine. The line is nonexistent, and the admission is 5 euros. If art is de-fined as "the expression or application of human creative skill and imagination" (*Oxford English Dictionary*), the world below contains some of the finest art ever created—art that has afforded our very survival. And like all great art, the more you understand the backstory, the more you appreciate what's before your eyes.

The museum is located in a working sewer, but like the Metropolitan Wastewater Treatment Plant, it smells no worse than a high school locker room on a damp day. As I descend a small flight of concrete stairs, I'm unsure whether I'm about to enter a Fellini movie or a Ken Burns documentary; turns out, it's both. I'm immediately taken by the complexity of it all. The sewers carry more than sewage; with brilliant foresight, the early designers sized the sewers to accommodate massive drinking and nondrinking water mains plus then-unimaginable things like control cables for traffic lights, fiber-optic cables, compressed air distribution pipes, telecommunication cables, and pneumatic tubes. (For safety reasons, electrical wiring and gas aren't included.) The engineers also designed the system to be large enough for workers and machinery to function comfortably—and to accommodate the throngs of curious visitors who, as early as 1889, began taking escorted tours through the sewers on boats and in wagons.

I stroll on walkways past vintage sewer flushing machines, mannequins of present-day sewer workers, dioramas of rats (once, but no longer, a major issue), and images of historic floods that accelerated the development of Paris's sewer system.

As I weave through the labyrinth of chambers, I discover they're marked with street signs that mirror the streets and buildings above for ease of navigation and maintenance. Wastewater churns through the troughs below, and smaller feeder sewers contribute their contents along the way. These are the sewers Victor Hugo described in his 1862 novel *Les Misérables*: "Paris has another Paris under herself; a Paris of sewers; which has its streets, its crossings, its squares, its blind alleys, its arteries, and its circulation, which is slime, minus the human form." This is the underground world

Waste from the city of Paris was dumped into drains that led directly into the Seine—the same river drinking water was drawn from.

COURTESY OF THE MUSÉE DES ÉGOUTS DE PARIS (PARIS SEWER MUSEUM)

where Jean Valjean descends into the "entrails of the monster" carrying the wounded Marius to escape detection by soldiers. Hugo finds it to be "a melancholy thing to be caught in this Paris of shadows." But I don't.

I turn the corner to enter the Belgrand Gallery, the heart of the museum. Here I find the history of sewers befittingly perched on a three-hundred-foot-long grate, covering a concrete river carrying some of the two million cubic meters of wastewater the nearby Achères treatment plant processes every day. The displays begin with the Roman era, when wastewater was poured onto fields, dumped onto unpaved streets, or tossed into the Seine. Around 1200, King Philip II had the main streets paved and incorporated an open gutter down the center for wastewater released into the Seine. But the putrid pools and quagmires that accumulated in other areas created conditions ripe for the black death. Fanned by fleas, rats, and wrong-headed medical superstition, the plague claimed as many as half of Europe's citizenry.

In 1370, a stone-walled sewer was built under Rue Montmartre. By the early 1500s, Francis I made cesspits mandatory under all buildings. The contents were removed by hand at night and emptied into surrounding moats and rubbish dumps by the Maître Fifi. During the reign of Louis XIV, a ring sewer was built on the Right Bank, and under Napoleon, twenty miles of vaulted sewers were built. While these sewers helped usher wastewater out of heavily populated (and mostly affluent) areas, more than one hundred thousand cubic meters of wastewater per day were being dumped directly into the Seine, the same river that supplied most of Paris's drinking water. The cholera epidemic of 1832—which claimed another nineteen thousand Parisians—was a clear sign a permanent solution needed to be developed.[3]

Finally, in 1850 hydraulics engineer Eugène Belgrand—the Steve Jobs of sewers—was charged with the monumental task of revamping and redesigning sewer *and* water systems. He quadrupled the capacity of the city's sewer system and, in digging up the streets to install the system, is partly responsible for Paris's wide, charming sidewalks and streets that we enjoy today. His system included actual treatment of the sewage before it was discharged far downstream from the urban center. He masterminded the whole enchilada—storm sewers to prevent flooding, aqueducts that doubled the water supply, and devices that used the power of the wastewater within the sewer to clean the sewer. The most ingenious of these were "cleaning balls," metal and wood balls slightly smaller than the diameters of the pipes behind which wastewater would build. This created pressurized "jets" along the perimeter to push sand and solid waste in front of the ball and, in the process, clean the pipe. Other major upgrades were implemented in 1935, 1970, and 1991.

The sewers in my hometown don't have the history or size or cleaning balls of Paris, but they do share the same complexity and onerous task. When I ask city engineer Shawn Sanders whether there's a simple way to explain the basic workings of our sewer system, he scrunches up his face and points to a shelf full of encyclopedia-size books. But he gives it a shot.

The sewage pipes leading out of most homes are four to six inches in diameter. They're installed with a slope of about a quarter inch per foot. This slope is critical: not enough slope and the solids flow too slowly or not at all; too much slope and the water runs away without carrying the solids with it. These pipes lead to *laterals*, or branch lines, in the street, usually eight to twelve inches in diameter, which lead to even larger submains and, eventually, the main sewer line. Pipes can be made of vitreous clay, cement, PVC, even brick. According to Sanders, the optimal flow rate in laterals is two to three feet per second—too slow, and you're facing a backup; too fast, and you have a serious mess on your hands. To achieve this rate, these pipes too need to be laid at the precise slope.

Every four hundred feet or so, the sewer lines enter below-street-level manholes, vertical concrete pipes that flare out to four

Early metal and wood sewer balls used for cleaning hard-to-access sections of the Paris sewer system. Versions of these are in use today.

or five feet in diameter. Some extend twenty feet or more downward. The manholes provide a place where workers can enter the sewer system to maintain the pipes or make repairs. The four hundred feet of spacing between manholes is based on the "reach" of the water jet and vacuum equipment used to clean the pipes. The manhole can also serve as a place for sewage to make a quick vertical drop.

Sewer lines increase in size as more and more pipes contribute their contents. These eventually feed into interceptor lines that lead to the sewage treatment plant. Ideally gravity—which never has a mechanical breakdown—delivers the sewage to its destination. But in low-lying or distant areas, lift stations and pumps are used to boost the sewage to a higher elevation so gravity can do its job.

"This is a ridiculously simplistic view of the process." Sanders smiles. "But it's better than no view at all."

WALK THE WALK

Ten Things You Should NEVER Flush

The manager of one sewage treatment plant explained to me, "Toilets aren't hazardous waste dumps or garbage bins. And when people use them as such it makes our jobs—and the equipment's jobs—harder." Here are ten things he'd rather never see again:

1. *Flushable baby wipes.* Toilet paper is made with short fibers that break down quickly. "Flushable" wipes have long fibers that don't, meaning they clog sewer pipes and bar screens.

2. *Dental floss.* It acts as a lasso, binding together debris that can clog sewer pipes and equipment.

3. *Kitty litter and dried poop.* There's not enough water in a flush to move a mass of litter through the system, and dried poop is basically petrified rock that doesn't break down.

4. *Bleach, paints, and solvents.* The treatment system isn't designed to remove toxic chemicals.

5. *Prescription and nonprescription drugs.* See No. 4 above.

6. *Cigarette butts and chewing gum.* They don't dissolve and need to be removed mechanically at the plant.

7. *Hair.* Like dental floss, hair can create a tangled mess that can clog pipes.

8. *Tampons and pads.* They don't disintegrate and must be mechanically removed.

9. *Fats, oils, and grease (FOG).* They can accumulate and clog even large sewer pipes, creating backups. A fifteen-ton "fatberg" was recently discovered clogging one city's system in England.

10. *Dead pets.* Goldfish, gerbils, and snakes don't decompose (at least not fast enough).

We climb back up through a manhole cover—that taken-
for-granted portal to the underground world designed to endure
the rigors of street life. Making one requires a surprising amount of
handwork: aluminum patterns are used to create sand castings;
scrap metal is heated to 2,700 degrees F and poured into molds;
hammers, vibrators, and brushes are used to remove excess metal
and slag; and the completed manhole is turned on a lathe so the
outer edge will rest flat on the circular metal frame. In the end, you
have a one-hundred- to three-hundred-pound disk. And the world
needs a lot of them. My little hometown of Stillwater has three
thousand just in its sanitary sewer system. Large cities—where
electricity and communication wires also travel belowground—
have hundreds of thousands.

All is not pure function in the world of manhole covers. Japan
has installed art underfoot depicting Monchhichi characters, dino-
saurs, and Ninjas.[4] Minneapolis has manhole covers celebrating
cultural diversity, loons, walleyes, and lady's slippers.

There's a quip: "Tragedy is stubbing your toe on a manhole
cover; comedy is watching someone else fall through an open one."
But it's not comical to the twelve-year-old from South Philly who
fell through an open manhole, winding up in the hospital. The
missing cover was one of six hundred stolen by thieves during the
economic downturn—and scrap-iron-price upturn—of 2008. A
two-hundred-pound cover could bring as much as $30 at the
scrapyard—paling in comparison to the $500 replacement cost—
but greed, not logic, ruled the day. Philadelphia is not alone; ten
thousand manhole covers a year are stolen in Bogotá, Colombia.
Police in Chicago recently tracked a van and witnessed three occupants stealing sixteen sewer grates in fifteen minutes. Some cities
have begun welding, bolting, or locking covers in place. One city
now embeds GPS tracking chips in its covers.[5]

Michael Clendenin, a spokesperson for Con Edison, which has more than two hundred thousand manholes in the New York City area alone, presumes most stolen covers wind up as scrap. "I can't imagine people are decorating their living rooms with them," he says.[6]

And why are they round? A round manhole cover can't fall through its own opening. A circular cover can be installed regardless of how you position it. It can be more easily moved by rolling. It mimics the shape of the person about to enter it. The cylindrical culverts below are the strongest shape in resisting compression. It takes up less surface area than a square cover of equal diameter.

Would Microsoft have hired you?

TRASH

How to Fit Three Tons of Trash into a One-Pint Jar

THE AVERAGE FAMILY OF FOUR IN THE UNITED STATES GENERATES 6,424 pounds of garbage per year. Bea Johnson's family generates half—not half of 6,424 pounds, but literally one-half pound of garbage—hardly enough to fill a mason jar.

When I ask Johnson whether it takes a lot of sacrifice to whittle down one's trash and lifestyle to the size of a mason jar, she responds, "Quite the opposite. We found it isn't just good for the environment, but for us too. It's made us healthier, it's saved us 40 percent of our overall budget, and it's made room in our lives for the things that matter most to us. Plus, our kids know how to use a handkerchief."

But come on—a pint jar? I generate that much trash at Chipotle for lunch.

There have been only two structures made by humans massive enough to be seen from outer space with the unaided eye: the Great Wall of China and the now-closed Fresh Kills Landfill of Staten Island, the resting place for fifty-three years of New York

City's solid waste.[1] It consists of four mounds of garbage; the tallest is 225 feet in height, 75 feet taller than the Statue of Liberty. "Give me . . . the wretched refuse of your teaming shore," indeed.[2]

Americans generate 250 million tons of trash every year—5 percent of the earth's population generating a quarter of its garbage. The good news is that about 60 million tons of that trash gets recycled, and 20 million tons of yard and kitchen waste gets composted. The bad news is that the remaining 170 million tons gets tossed.[3]

If I were to kidnap your week's worth of trash on garbage day and pick through the fifty pounds of waste left *after* you'd separated out the conventional recycling, I'd still find:

- 10 ½ pounds of food waste
- 9 pounds of plastic
- 7 ½ pounds of paper and cardboard
- 5 ½ pounds of rubber, leather, and textiles
- 4 ½ pounds of metal
- 4 ⅓ pounds of yard trimmings
- 4 pounds of wood
- 2 ½ pounds of glass
- 2 pounds of "other stuff"

If you sort through the above list, you'll find that about 80 percent of it could be recycled or composted. If you were to sort through Johnson's mason jar, you'd find several pieces of bubble gum, an expired laminated ID card, backing from postage stamps, masking tape from a paint job, and little more.[4] Why does our pile of trash dwarf Johnson's? Well, it turns out we have a long, sordid relationship with trash.

For 95 percent of human history, garbage consisted of ashes, food scraps, and human waste. All were naturally returned to the

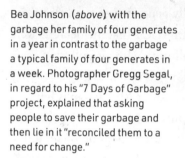

Bea Johnson (*above*) with the garbage her family of four generates in a year in contrast to the garbage a typical family of four generates in a week. Photographer Gregg Segal, in regard to his "7 Days of Garbage" project, explained that asking people to save their garbage and then lie in it "reconciled them to a need for change."

GREGG SEGAL, "7 DAYS OF GARBAGE" SERIES

earth. As populations grew, however, so did the trash problem. Most trash stayed close to home—sometimes just out the window. One archaeologist estimated that so much trash was deposited within the city of Troy, its elevation rose five feet every century. In the 1300s, so much garbage accumulated on the outskirts of Paris that city defenders complained it was difficult to spot invaders.

During the 1400s, the black death, caused by fleas and rats that reveled in the filth of the streets, wiped out 30 to 60 percent of the European populace. Though a firm connection between waste and public health wasn't established until the 1900s, some cities began dealing with waste.

In the 1750s, Londoners began throwing trash into the Thames, Parisians into the Seine. New Yorkers tossed theirs into the East

River. Many cities employed a "waste to swine" strategy, which let loose pigs to clean up street garbage. Because seventy-five pigs could consume a ton of food waste per day, piggeries became popular. In its heyday, Los Angeles sent five hundred tons of garbage daily to Fontana Farms with its herd of sixty thousand garbage-processing pigs.[5]

In the mid-nineteenth century, yellow fever and cholera accelerated the demand for better sanitation. Around 1875, the first incinerator, known as a "destructor," was built in England. Incinerators proliferated. Many eventually generated enough electricity to power the plant. Yet it soon became clear that incinerators were simply replacing land and water pollution with air pollution.

The popularity of dumps and filling in wetlands grew.[6] It surely wasn't a new concept; fully a third of lower Manhattan is built on fill. Some merchants knowingly purchased lots *under* the East River, filled them with rubble and earth, and built on them; today, this Water Street area sits two blocks inland. At the original World Trade Center site, which was formed largely from earth removed to build the Lexington Avenue subway, excavators found an eighteenth-century shipwreck, an African burial ground, and endless debris from burned or demolished buildings.[7]

In 1895, George Waring—the rock star of garbage management—was hired as the commissioner of the Department of Street Cleaning in New York City. He imposed a military-like order, creating the White Wings—white-suited sanitation workers—who transformed the city. He established a form of recycling in which households were required to separate ash, food waste, and dry trash into three separate bundles or bins. Albeit short-lived, his model became the one many other cities followed.

The volume of trash and methods for collecting it increased in step with new types of packaging. In 1795, Napoleon offered a 12,000-franc reward to anyone devising a way to preserve food for his military forces. Fifteen years later, Nicolas Appert claimed the prize by inventing the can. Cans weren't used just for food, but for gunpowder, seeds, and turpentine as well. Today, Americans crank open 130 billion steel cans a year.[8]

Cardboard and corrugated cardboard boxes eased onto the scene during the 1850s. Cereal boxes appeared around 1900. In 1904, the automatic bottle machine was invented. By 1919, twenty-two million glass containers were being made in the United States. A century later that number skyrocketed to forty billion.[9] In 1954, Dow Chemical introduced Styrofoam. Three years later, the first Styrofoam cup debuted; today, twenty-five billion Styrofoam cups are tossed per year.

The growing popularity of television whipped consumers into a buying frenzy. If you didn't buy Dixie cups, disposable diapers, and boxed cake mix, you were an absolute troglodyte. In the late 1950s, polyethylene milk jugs, TV dinners, BIC pens, and disposable razors marched into stores.

In the 1960s, the single-use plastic shopping bag was created. About five trillion are used per year worldwide. Americans use seven hundred per year, Danish shoppers, four.[10]

Plastic clamshell packaging became popular; products were visible on both sides, could be hung on peg hooks, and made shoplifting more difficult. The advent of single-use coffee servings has led to fifty-six billion coffee pods and capsules being used annually, though only about 25 percent are ever recycled. Some are being reborn as RE:CYCLE bicycles—complete with coffee cup holders—each made of three hundred used aluminum Nespresso pods.[11]

Items are now packaged inside of packages inside of packages. "When you buy most things, at least 15 percent of the cost is in packaging you throw away," Johnson bemoans. "It's a waste of dollars and resources."

To contain all this giddy new packaging, new types of containers and collection emerged. Steel trash cans were replaced by plastic. Glad garbage bags appeared on the scene. In the 1960s, automated cart systems, with wheeled carts and claw-wielding trucks for dumping them, were introduced.

With bottomless resources and a more-is-better mentality, the United States created more trash than any other country. How to dispose of it became a challenge—and the bigger the city, the bigger the problem. Stories like the 1987 Mobro barge incident, where the 230-foot barge, loaded eighteen feet high with garbage, wandered for five months before finding a place to relieve itself, became common.[12]

Bea Johnson, where are you?

Though she was born in France, Johnson's English is impeccable— and the lingering French lilt disarming. If I were given two words to describe her, they would be "sincere" and "classy" in no particular order. She has the ease and confidence of a woman who just might try to persuade the world to reinvent its way of dealing with its trash and—as someone who's given hundreds of talks in sixty-five countries—is doing just that.

Though dubbed the "Priestess of Waste-Free Living," her ascent to the throne was littered with twists and turns. In her book _Zero Waste Home_ she explains, "We had no financial worries, as life rolled by effortlessly and afforded my Barbie-like platinum-blonde hair, artificial tan, injected lips, and Botoxed forehead. . . . We seemed to have it all."[13]

But Johnson and family discovered they had too much "all."

The first step to zero waste was to simplify their lives. They moved from a three-thousand-square-foot home in a "drive-only" San Francisco suburb to Mill Valley, north of the city and within walking distance of shops, cafés, libraries, and hiking trails. While in transition, they put most of their belongings in storage. By the time they moved to a house half the size of their former one, they realized they didn't need most of the stuff in storage and got rid of 80 percent of it. She and her husband began reducing energy consumption and waste while increasing their awareness of what kind of world they, as parents, were leaving for their kids. By simplifying, they saved enough money to install solar energy panels and a gray water system that uses bath water to irrigate the garden.

"When I began, there wasn't much information on how to reduce trash," Johnson reflects. "I had to pick up the phone and ask my mom, mother-in-law, and grandmother how the heck they did it." Johnson admits she went overboard. In her effort to reduce packaging and waste, she tried using moss for toilet paper and stinging nettle as a lip plumper. She cut her hair to reduce the amount of hair products she used. But, in the end, she realized their family needed to make changes that were feasible in the long run. They had to "find a sustainable balance between being easy and being green."[14]

Most of us have heard the alliterative mantra Reduce, Reuse, Recycle. But Johnson has bookended this with two of her own *R*s: on the leading end is "Refuse" (not as in "rubbish" but rather as in "no thank you"), and on the tail end is "Rot" (as in composting). She applauds the many communities now offering curbside organic waste pickup and explains, for areas that don't, that there are scads of composters on the market. "Worm composters are a great illustration of the cycle of life if you have kids," she explains. "Some look like stools, so you can keep them around your kitchen table."

When I ask Johnson how US garbage compares with that in other countries, she tells me, "Consumerism has been in the US longer than in other parts of the world, so in some places—especially Eastern Europe—it's easier for people to live simply because they still remember when they lived simply. They still know what a handkerchief is." She explains Europeans are more accustomed to buying meat from the butcher and bread from the baker wrapped in paper. They're more likely to buy fresh fruits and vegetables from an open market. She's a huge fan of bulk buying, which allows people to bring their own bags, jars, or bottles to stores, thus cutting down on package waste. Her website Zero Waste Home and its app list forty-five thousand sources for buying in bulk. "Most people aren't aware of bulk products," she says, "but once your radar goes up, you see them everywhere."

But few of us buy in bulk, keep worms in our kitchen, or use stinging nettle for lip plumper. We toss. It's called municipal solid waste (MSW), which either goes up in smoke or gets buried. In a strange twist, energy is increasingly being derived from both options.

After composting and recycling, about half our garbage heads to one of the two thousand MSW landfills in the United States. Between 1960 and 2015, the amount of MSW tripled; thanks to recycling, that rate has since slowed.[15] Unlike the dumps of yore, where trash was simply dumped, landfills are highly engineered. If you wished to build your own landfill, you'd first need to tackle a mountain of red tape and then dig a massive hole and create a six-foot-thick liner system consisting of two feet of compacted clay, a beefy sixty-milliliter plastic membrane, a leachate collection system with drains and pipes, a foot of fill, another six inches of clay, and another geomembrane, all topped off with another leachate collection system.[16] You'd also need to install a pipeline system for collecting methane and delivering it to a power plant or other facility. The company Waste Management uses this methane to power 6,536 trash-collecting trucks. Only then can you start layering in trash. When you've filled it up, be prepared to install a sixty-inch-thick cover of equal complexity. Then turn it into a park.

A smaller percentage of US garbage gets burned in one of eighty-six waste-to-energy (WTE) facilities.[17] There, garbage is stockpiled in warehouses and then dumped onto moving grates using colossal overhead claws. Air and fuel are added to increase temperatures to at least 1,560 degrees F, hot enough to break down toxic chemicals. The resultant ash—about 10 percent in volume compared with the garbage from which it came—is sifted and sorted for remaining metals.[18] It is then used in construction materials or as nonstructural fill, or it heads to the landfill.

It's estimated that WTE plants generate enough electricity to power two million homes in the United States.[19] The CopenHill incinerator in Denmark not only converts half a million tons of waste into energy to electrify and heat 150,000 homes annually, but it also serves as a recreational area complete with a quarter-mile year-round ski hill, a 275-foot climbing wall, gardens, trees, and hiking trails on its roof.

The WTE process is not without its critics. Many say it lulls people into recycling less, knowing that their paper, cardboard, plastic, and yard and kitchen waste will be burned rather than dumped. Others say relying on WTE power detracts from more environmentally friendly options such as wind and solar.

More than seven thousand cities have adopted pay-as-you-throw (PAYT) or garbage-metering programs that involve charging varied rates for different size bins, rather than charging a flat rate. Some communities sell specially marked bags ($2 per thirty-three-gallon bag is relatively common). The city of Brewer in Maine (population about ten thousand) saw a 50 percent reduction in waste, a 50 percent increase in recycling, and a savings of $370,000 after instituting PAYT.[20] In San Francisco, residents expect to pay $27.50 for a sixty-four-gallon bin of trash, but only $13.74 for their sixty-four-gallon bin of compost or recycling.[21] Eighty percent of the city's trash is diverted from landfills, compared with

10 percent in Chicago.[22] Clearly, one effective way to impact trash is to impact wallets.

Our trash collector's name is Nick; he commandeers a truck that weighs twenty-five tons fully loaded and costs about a quarter million dollars. It holds eighteen thousand pounds of compacted trash, enough for a full day of collecting before he heads to the transfer station or landfill.

Nick most likely makes $40,000 to $70,000 a year, though some collectors make six figures.[23] There's a two-thirds chance your trash collector likes his or her job, but only a 1 percent chance he or she is a "she."[24] In some cities, competition for the job is fierce; if you apply for a job at the Department of Sanitation in New York City, you have a one-half of 1 percent chance of being hired.[25] If you apply for enrollment at Columbia University down the block, your chances of getting in are twelve times greater. In some cities, your trash collector (and truck) also plow your snow.

Robin Nagle, author, female sanitation worker, and professor at New York University, explains that due to the "mundane, constant, and largely successful nature of [the trash collector's] work," they essentially become invisible to many people. "His truck and his muscle punctuate the rhythms of a neighborhood at such regular intervals that he becomes a kind of informal time piece"—someone taken for granted and not much more.[26]

Trash collectors work in one of *the* most dangerous professions. They're more likely to be killed in the line of duty than a police officer. The biggest danger is getting hit by a passing vehicle. Knees, backs, legs, and necks take a beating. Despite wearing heavy gloves and clothing, workers get cut by broken glass; stabbed by hypodermic needles; clobbered by bowling balls launched from hydraulic compactors; bitten by rats, cats, and raccoons; and inhale toxic fumes. In 1996, Michael Hanly was doused in hydrofluoric acid while compacting a load of garbage. Two thousand sanitation workers showed up at his funeral.[27]

Treasures in the Trash Museum
ONE CITY'S WASTE IS ONE MAN'S PASSION

After thirty-four years of slinging thirty-pound bags of trash into the hydraulic jaws of a garbage truck, most people would have a bad back and bad attitude—but not Nelson Molina. He's got a museum with more than forty-five thousand treasures he's rescued from the trash and the disposition of a motivational speaker.

I'm standing with this sixty-five-year-old icon of New York City sanitation on the top level of the "Man 11" truck garage— home to his Treasures in the Trash Museum. "Every time I come here," Molina says, "I ask myself how I ever put all of this together. I arranged everything, hung everything, I scraped the walls and painted them. If I had a few minutes to kill after a shift, I might work up here, but mostly I've done this on my own time."

His collecting passion kicked in at an early age. "When I was nine, ten, a couple of weeks before Christmas I'd scour the area around our apartment. I'd find busted toys and fix them up. If a car was missing a wheel, I'd replace it with a button. Then I'd give them to my brothers and sisters." He inherited his fixer-upper aptitude; he remembers his mother fixing their toaster by holding a butter knife over a stove gas burner and then using it as a soldering iron.

After spending time with Molina, you're forced to trash every preconceived notion you might have about a sanitation worker. He runs triathlons, works out three times a week, and looks it. "Every time I go to my doctor, he says I have the heartbeat of a

Nelson Molina unsheathes a Samurai sword—one of forty-five thousand items he rescued from the trash during his thirty-four years as a sanitation worker—now part of his Treasures in the Trash Museum.

twelve-year-old," he grins. He's won humanitarian awards from the Asian Jade Society and Hispanic Society and has appeared in publications ranging from *Country Living* to the *New York Post*.

He's unassuming and unapologetic. His son, two brothers-in-law, and a godson all work in the facility below. When I ask whether a third generation could be working there someday, Molina rubs his hands together and replies, "I can only hope so."

Like all curators, Molina has his treasures arranged in themes—on tables and shelves, two hundred of them, all rescued from the trash. We stroll past tables dedicated to Superman dolls, African carvings, Buddhas, Furbies, typewriters, PEZ dispensers, and stringed instruments.

He has a sixth sense. "Even in the bag I can hear the difference between the rattle of a Coke bottle, a wineglass, and a vase," he says. "Or if a building usually has five bags and one week there are ten, that tells a story. If you give me three months, I could furnish a three-bedroom apartment, no problem."

Molina's beat was the mostly affluent area from 96th to 110th Streets, between First and Fifth Avenues; his treasures are wide-ranging. There are vintage Steiff bears with tags intact, autographed photos of Jackie Kennedy, taxi meters, twelve-foot papier-mâché moons, Fender guitars, 125-year-old stained-glass windows pulled from a church trash pile, and enough exercise equipment to set up a gym (which he has, for those who work in the garage below).

Halfway through the museum he pauses to plop a vintage disk into a vintage CD player; we're serenaded by Dean Martin as we walk through history. We examine metal social security cards, a brass commemorative bookend from France ("I wish I could find the other one"), a banner from the never-held 2012 New York City Marathon, promotional literature from the *Queen Mary*, a signed book by Lena Horne, and a jewelry box that plays "Love Me Tender."

The occupation has its thrills and chills. If Molina found a knife lying atop a trash bin, he had a potential crime scene on his hands. "One time we found a rolled-up rug with blood coming out the ends. I'm not touching that," Molina says. "We call our supervisor, he calls the cops, they rope off the area, unroll the rug—and it's a deer."

Molina stops before a small display case, reverently opens the lid, and carefully plucks out a Star of David the size of a hand. "This is dear to me," he explains. "It's cut from a beam from the World Trade Center. September 11 was a Tuesday. I volunteered to help with the cleanup, and on Sunday I'm working there with a paper mask, pushing dust all over the place. I get a call from my wife saying my father was just rushed to the hospital. Three days later he passed. We buried him and I wanted to go back to help, but so many people volunteered they couldn't hold a spot for me. That's the sad part. But it's also a happy story because, if I'd stayed there working for a year or two, I may not be here today. Seventeen, eighteen years later guys working down there are still dying from inhaling all that poison."

New Yorkers generate about twenty-five thousand tons of garbage every day—most of it shipped to landfills, some as far away as South Carolina. When I ask Molina what people can do to reduce the amount of their trash, he tells me, "At least 50 percent of garbage isn't garbage. People have to recycle. They have to bring stuff to a thrift shop or Goodwill." He practices what he preaches. "In my building I set up a table in the basement, and if there's something you don't want, you put it there. I just put a toaster oven out and someone took it. I read the *Daily News* every day, and when I'm done I put it on the table; when people are doing their laundry they'll read it and put it back and someone else reads it."

We walk past a five-foot sailfish, Picasso posters, autographed baseballs, Cabbage Patch Dolls, and yo-yos. "I know everything I have in here and I know if someone moves something," he says. When I ask him whether he ever thought he'd be a museum curator, he laughs. "When I began I didn't even know what a curator was. But now I'm a curator." And those at the Guggenheim a dozen blocks away don't have anything over Molina.

Note: The Treasures in the Trash Museum isn't open to the general public. At the time of publication, tours could be arranged through the New York Adventure Club (https://www.nyadventure club.com).

ROADKILL (AND LITTER)

Squished, Plucked, *and* Plogged

WHEN MOST OF US SEE A DEAD POSSUM OR SKUNK ON THE SIDE of the road, we avert our eyes, swerve to the left, pinch our nose, or all three. Not Heather Montgomery. She hits the brakes, pulls over to the shoulder, dons her latex gloves, and then . . . well, she does something you and I would never do.

The argument can be made that if a critter is so clueless as to think it's fast enough or tough enough to stand its ground against a Dodge Ram, it deserves to become URP—unidentified road pizza. Unfortunately, for the critter, we're often the clueless ones. The roads we build crisscross the paths critters use to migrate, mate, and seek out food, water, and shelter. The road salt we spread and apple cores we toss out the car window lure them like moths to a flame. Our distracted driving turns our vehicles into two-ton bullets.

Our collisions with deer, moose, caribou, and other large animals reduce their numbers by about one-and-a-quarter million a year—and reduce our numbers by about two hundred a year too.[1] Nationwide, chances are 1 in 170 that you'll file an animal collision insurance claim this year (1 in 46 if you live in West Virginia; 1 in 6,380 if you live in Hawaii).[2] Insurance companies shell out

Heather Montgomery had always wondered whether she could skin a skunk without releasing its "essence." She found she couldn't, but gained respect for the natural defenses of the critter.

about $4 billion annually for animal collisions—and we shell out an equal amount in lost wages and medical expenses.

Roadkill differs from state to state and country to country. Armadillos and possums pose issues in the Deep South; it's moose and caribou in Alaska, deer almost everywhere, kangaroos in Australia, penguins in New Zealand, monkeys in Costa Rica, and crabs on Christmas Island.

What can we do besides avert, swerve, and pinch? Plenty.

I'm sitting with Montgomery in her home, overlooking the rolling hills of Ardmore, Alabama. Well, I'm sitting—she's zipping around. Because she's ecstatic about roadkill. She has the wide-eyed enthusiasm of the elementary school kids she teaches visit after visit, week after week. She's the author of *Something Rotten: A Fresh Look at Roadkill* and *How Rude! 10 Real Bugs Who Won't Mind Their Manners.* The title of her upcoming book is *Who Gives a Poop? Surprising Science from One End to the Other.*

She produces a two-gallon ziplock bag containing the complete skeleton of a fox, followed by pelts and a plastic box holding feathers, skulls, and tails. We examine deer vertebrae through magnifying loops. With no small amount of glee, Montgomery shows me the skinning knife her husband gave her as an anniversary present. She hands me a skunk pelt to fondle. "I had to sleep on the porch the night I skinned that one," she discloses.

And while she makes roadkill engaging—even fun—she stresses there are lots of simple things we can do to reduce it. Deer, elk, and many other animals are *crepuscular*, meaning they're most active around dusk and dawn. They're also particularly active in the fall during mating season when love is in the air; indeed, your chances of hitting a deer double in November. "The number one thing you can do is to adjust the time of day you drive. If you can't do that, slow down," Montgomery explains. "And if you pay attention, you'll begin noticing roadkill hot spots." Waterways, trash sites, or wooded outcroppings where animals hide before dashing across the road are notorious death zones. "And don't dump stuff out the window. Emptying a can of Coke or tossing a banana peel on the median may seem harmless, but that's food. It's inviting animals to the road. It might be only a mouse, but that mouse attracts foxes and owls.

"AND," and this is where we find Montgomery pulling on the latex gloves at the beginning of this chapter, "it's critical we get dead stuff off the road. Roadkill attracts other animals that then become roadkill." Montgomery salvages some roadkill to use in her presentations, but she drags most to the ditch. "I don't do this as much as I used to, because I'd find so much roadkill I'd never get to where I was going. I made a rule for myself: I could only stop on my way home." She drags off a lot of armadillos. "They're slow, have lousy eyesight, like to travel along the mown shoulders of the road, and have this crazy adaptation that makes them jump straight up when a predator approaches. That might scare off a dog or a coyote, but it puts them right on the bumper of a car."

Why are some animals more prone to becoming roadkill than others? "With rabbits, possums, squirrels, or other preyed-upon animals, their self-preservation instinct is to freeze when in danger; freeze and the predator, which normally detects them through motion, might not see them. They see cars as predators." Deer are particularly vulnerable because of their eye anatomy; their eyes adjust well to low light conditions, but a high beam hitting those eyes at night literally blinds them. And they act like a deer in the headlights.

We have our corridors—freeways, highways, and railways—and animals have theirs; unfortunately, the two often intersect. That's part of the rationale behind an $87 million wildlife crossing being built over Highway 101 north of Los Angeles. Its purpose is to provide cougars, deer, coyotes, lizards, snakes, and other animals safe passage to adjacent open spaces where they can find more food, elbowroom, and prospective mates. This will minimize roadkill while maximizing the breeding pool for mountain lions, like the one nicknamed P-22 who's been marooned by crossroads in LA's massive Griffith Park. Beth Pratt of the National Wildlife Federation explains that cougars "can't get out of here to get dates, and cats can't get in to get dates." For people who live in LA, she quips, "Having a romance prospect quashed by traffic is something we can all relate to."[3] But $87 million? It depends on the value you assign to the lives of drivers, animals, and perhaps even the survival of a species.

There are thousands of wildlife bridges and tunnels around the world. On Christmas Island in the Indian Ocean, metal bridges allow fifty million crabs to move unsquished from their homes in the rain forest to their spawning grounds in the ocean.[4] Crossings in Mexico save jaguars. Underpasses in Africa save elephants.

Some measures to reduce roadkill are counterintuitive—like the reintroduction of cougars and wild cats to reduce the number of deer. When they did this in South Dakota, deer hits declined, and insurance claims actually dropped.

Our roads will never be free of roadkill. Some cities and groups are making the best of it. In southern Alabama, roadkill deer are fed to the reptiles of Alligator Alley, a sanctuary for nuisance gators. In New York—a state that needs to dispose of twenty-five thousand deer carcasses a year but can no longer landfill them because of chronic wasting disease (a transmittable, always fatal disease loosely related to mad cow disease)—the Department of Transportation has been creating compost. The recipe? Dump one truckload of woodchips gleaned from roadside tree trimmings on the ground, add four deer carcasses, cover with more woodchips; repeat. Wait as nature bakes at 160 degrees F. Remove teeth and large femur bones. Yields nutrient-rich soil that can be used for state landscaping projects.

In Washington state, where you can recover roadkill with a permit, sixteen hundred deer and elk were picked up and processed in a recent year. Similar laws exist in my home state of Minnesota and nineteen others. "If you hit a deer and you've got a car with $5,000 damage, the least you could do is get a little meat in your refrigerator," explains Jay Kehne, of the Washington Department of Fish and Wildlife.[5]

In Alaska, where six hundred to eight hundred moose are killed by cars each year—each with two hundred pounds of organic, free-range meat—the law requires you to report a hit. State troopers contact charitable organizations that have volunteers who take the moose to a butcher or dress it themselves. The meat is donated to people in need.

And then there are birds of prey; a roadkill deer presents a veritable all-you-can-eat buffet for them. Fully engorged eagles have trouble getting airborne until their food digests, so they can become roadkill themselves. The National Eagle Repository is a clearinghouse for carcasses and feathers, regularly receiving dead eagles—on ice—from zoos, state departments of natural resources, and individuals. Native Americans wanting feathers or "eagle parts" for ceremonial or religious purposes can make requests of the repository, though obtaining the "whole tail" of an immature golden eagle can involve a wait of up to five years.[6]

And then there's seventy-seven-year-old taxidermist Arthur Boyt of England who's been eating fresh roadkill since he was thirteen. He touts the benefits: it's organic, it's hormone- and antibiotic-free, it gets bodies off the road, and the price is right. "I've been ill from eating food from a buffet," Boyt explains. "But I've never ever been ill from eating roadkill."[7]

Nor have I.

I had the chance to vicariously walk in Boyt's footsteps as a judge in 2019 at the annual West Virginia Roadkill Cook-Off in Marlinton. According to the rules, "All entries must have, as their featured ingredient, any animal commonly found dead on the side of the road—groundhog, opossum, deer, rabbit, bear, crow, squirrel, snake, turkey, etc." In past years, iguana, crawdads, and alligator were served.

I had imagined downing some nose-holding fare akin to the warthog anus or beating python heart Anthony Bourdain had eaten for his travel and food show *Anthony Bourdain: Parts Unknown*. But I was wrong. Eight of the eight dishes were flavorful and well-presented (most had been recently hunted). I downed "Bad News Bears Bumper Tacos," rabbit gumbo, "Side Ditch Doe," and more. The unanimous winner was wild boar served in

Wild game expert Kevin Fraser (*left*), chef Tim Urbanic (*center*), and the author (*right*) sample the "Side Ditch Doe" while judging the West Virginia Roadkill Cook-Off. A wild boar dish was declared the winner.

stuffed poblano peppers drizzled with cheese, which made me realize how much I had missed by limiting myself to the Big Four: beef, chicken, fish, and pork.

Montgomery writes about roadkill being used by medical researchers to study cancer and by scientists to learn more about fossils. She talks about kids in Arizona who invented a solar-powered motion sensor to warn drivers of wildlife near the road and of a Massachusetts woman who founded Peace Fur using "accidental fur" to create fox, bear, and raccoon neck muffs. Montgomery waxes about how excited kids get when they handle the ball and socket joint of a fox and realize 'it's the same ball and socket as their own. And, in the end, you realize, the study of roadkill isn't about death; it's about life.

HACKS & FACTS ## Plogging and Pliking Picker-Uppers

If you've ever picked up a bottle, bag, or butt while hiking around the block, you're not just a walker, you're a *pliker*. The term—created by combining "hiker" and *plocka upp*, the Swedish term for "picking up"—involves picking up litter while hiking or biking.[8] (If you do the same while jogging, you're a *plogger*.) It's not just good for the environment, it's good for you. One fitness expert determined that people stopping, stooping, and starting while pliking can burn one hundred more calories per hour than those who merely walk.

There's plenty of pliking to be done, with an average of 6,279 pieces of litter per mile of roadway. The good news? The overall litter rate has actually been halved over the past forty years.

PART III

SURFACES

BIKE LANES

Pedaling Uphill
and into the Wind

I CONSIDER MYSELF TO BE IN DECENT SHAPE—EVEN WITHOUT THE disclaimer "for a man my age." Yet while I was cycling Brown's Creek Trail the other day, a discombobulating thing happened. I was moving along at a good clip when I glanced in my rearview mirror to see a cyclist advancing at an even better clip. The approaching cyclist was older, had fewer gears, and was wearing way less Lycra than me. My testosterone did not like this one little bit.

Determined not to be passed, I picked up the pace—yet my rival continued to gain. I rejoiced at seeing a hill ahead, knowing it would provide an opportunity for me to crush my unwitting opponent. Thirty seconds later, the chubby old fart passed me. Swiftly. Smiling. Not breaking a sweat. I gave him that little biker head nod and wheezed on, but not before craning my neck to read his bike decal: E-BIKE—the development destined to change not only my facial expression but the face of biking.

Given that the wheel evolved fifty-five hundred years ago, one wonders why the two-wheeled bike—not a far cry from the two-wheeled chariot developed in 3000 BCE—was so late to the party.

Eventually, the first bikes were designed—these as "running machines," dual-wheeled affairs with a seat in the middle so the rider could sit while propelling the "dandy horse" by foot. Some attribute its development to Baron Karl von Drais, who couldn't find a horse to ride during Germany's "year without a summer" in 1816, when the country's equine population starved to death following massive crop failures.

Until the 1860s, when French inventors began adding pedals to the front wheel, early wooden velocipedes remained little more than novelties. The only way to gain speed, balance, and momentum with pedaled versions was to increase the size of the front wheel. The resulting "penny farthing" bike—with its massive fifty-two-inch front wheel and dinky eighteen-inch back wheel— became the rage. Only the boldest dared ride one, however, because the contortions required for mounting and dismounting those "boneshakers" were so extreme and because British roads were so poorly maintained.

In 1880, the "safety bike" was invented. It featured equal-size wheels and centrally located pedals with a chain to power the back wheel. The bikes were easier to operate, since cyclists could ride in an upright position, and to smooth the ride, some safety bikes featured pneumatic tires and brakes so one could actually stop. The Raleigh Company, founded in 1888 in Nottingham, England, became the world's largest bike manufacturer, soon cranking out more than two million a year. The craze was on; by 1895 more than three hundred bike manufacturers were in the United States alone.

Bikes were a game changer. They led to the "rational dress" movement, when women traded in their corsets and long dresses for bloomers and shorter dresses, which allowed them to pedal more easily. Susan B. Anthony praised bikes, maintaining they'd "done more to emancipate women than any one thing in the world." Doctors called them the biggest boon in physical exercise in two hundred years.[1] One biologist credited the bike as one of the most important weapons in combating genetic disorders, because the gene pool became wider as an unintended consequence of suitors traveling farther distances from their homes in search of mates.[2]

The biking craze prompted improvements to local roads. Dedicated biking paths flourished. In 1894 the five-and-a-half-mile Coney Island Cycle Path opened, with ten thousand riders taking advantage of it on its first day. This path, which still exists today, was widened two years later to accommodate the growing number of cyclists. Other "sidepaths" proliferated. In 1900 the one-and-a-half-mile California Cycleway in LA was built. A one-way trip cost riders 10 cents, which let them enjoy a safe path where "there are no horses to avoid, no trains or trolley-cars, no stray dogs or wandering children." It's worth noting that many of the early sidepaths were "protected bike lanes" with barriers, fences, or landscaping separating them from horse and carriage traffic.[3]

When mass-produced automobiles roared onto the scene in the 1910s, road infrastructure changed with equal swiftness, especially in the United States. New roads were designed and built to accommodate the new horseless carriages, a national trend that never back pedaled. In the US, gas was cheap, the nation was expanding, and car was king. The bike movement came to a near-screeching halt. By the 1940s, the bike devolved into little more than a child's toy. By 1956 children and teens accounted for 90 percent of all bike riders.[4] It seemed the only adults who rode bikes were circus clowns and professional cyclists wearing jaunty little caps.

Meanwhile, across the Atlantic, the bicycle kept its balance and momentum. Old cities with narrower roads designed for foot and horse traffic were better suited for bikes than cars, and the density of European cities meant cyclists could easily cover the short distances between destinations. Some countries, like the Netherlands, were flat and easily bikeable.

Perhaps the largest reason bike infrastructure flourished on one side of the Atlantic but withered on the other was attitude. In the United States, the bike was used primarily for exercise and entertainment, while in Europe it was a model of convenience. Author Peter Walker observes: "The big changes happen when a nation doesn't see cycling as a hobby, a sport, a mission, let alone a way of life. They happen when it becomes nothing more than a convenient, quick, cheap way of getting about, with the unintended bonus being the fact that you get some exercise in the process."[5] Cycling can even be lifesaving; biking and bike-share program usage exploded (by as much as two-thirds in some cities[6]) during the coronavirus pandemic of 2020, as people sought out safer alternatives to commuting via crowded subways, buses, and carpools.

Are our cities too far behind the breakaway pack to catch up?
Not if we draft behind the cyclists of Montreal. In its early days, that island city was as inhospitable to bikers as the Sahara is to fish. There were few bike paths or bike parking places, an antibicyclist mayor was in office, and the primary way to get into, out of, or around the city was via cars or mass transit. Five bridges connected Montreal to the highly populated South Shore, and none was bikeable.

Because conventional attempts to create safe bike lanes through legal channels failed, an ad hoc group of cyclists called Le Monde à bicyclette, founded in 1975, took a different route. "We engaged in cyclodrama," Robert "Bicycle Bob" Silverman, the head of the bicycle collective, told me. "To highlight the absurdity of not being allowed to carry bikes on subways, we brought ladders, a fake hippo, and other large objects on the train."

Within a few days, the regulation was changed.

To champion bike lanes, Le Monde sponsored a race through the city between a bus, a subway, a car, and a bike; the bike finished 20 percent faster than any other mode of transportation. Le Monde organized a mass cycling parade, which drew three thousand riders. "It was one of the happiest days of my life," Silverman says, with a still-passionate lilt to his voice. By night, the group painted bike lanes onto city streets. They sponsored die-ins with ketchup "blood" to highlight the seriousness of the situation.[7] To emphasize the lack of safe bridges for bicyclists to cross the St. Lawrence River, one biking advocate dressed up as Moses in order to part the waters. One miracle did occur: the media loved it, and the movement grew.

Slowly the scene improved. A dedicated bike and pedestrian bridge was built in 1990. The number of bike lanes grew, and finally, finally, in 2017 Montreal pedaled into twentieth place on Lonely Planet's "20 Best Cities in the World for Cycling," the sole North American city to do so.

To catch a glimpse of the present and future, I talk with Gabrielle Anctil, a bicycle activist and journalist with Radio Canada. She is ecstatic. We're chatting minutes after she covered a press conference announcing the government's plans to add twenty-six kilometers of protected bike lanes over the next two years. "Biking has become so popular that we've reached maximum capacity on our current bike paths," Anctil explains. "Many of the existing paths are so narrow that it's impossible to pass, and we're seeing bicycle traffic jams."

Bear in mind, this is Montreal, a city where there's snow on the ground five months of the year and temperatures regularly dip to 10 degrees below zero. None of this has halted Montreal's bike scene from cranking full speed ahead.

"When I started biking to work in the winter a dozen years ago, my attitude was 'I'm going to bike in any weather, whatsoever, no matter what,'" Anctil says. "But now it's not a religion. Today, I bike 95 percent of the time, and if the weather is terrible, I take mass transit. When I bike in winter, I have a different attitude. I put on my ski goggles, ride an old single-speed bike, and think, 'I'll get there when I get there; enjoy the ride.'" She adds, "Bike lanes are usually cleared faster and are safer than sidewalks in winter anyway."

One of the memorial ghost bikes Anctil and other activists have installed at the location of fatal bike accidents to commemorate people killed on bikes.

In 2013, she and a co-worker started Montreal's Ghost Bike Project, where painted-white memorial bikes are installed throughout the city at sites of fatalities to call attention to cyclists killed by automobiles. The impetus was her co-worker arriving at work in shock one morning; the cyclist in front of her the night before had been killed when she swerved to avoid a car door flung open in front of her and was then hit by a bus.

When asked whether ghost bikes might give people the wrong message—that biking is dangerous—Anctil responds, "We put up fewer ghost bikes every year; they're a physical reminder that things are getting better. Every cyclist has had close calls—and it's a place to think about those close calls. There's a critical mass effect with biking. At ceremonies, we tell people if you want to make biking safer, bike more."

Fear of being hit by a vehicle is the number one reason people stay OFF bicycles. People aren't afraid of tipping over, colliding with other bikes or pedestrians, or having a heart attack; they're afraid of getting killed or injured by a vehicle. And in the United States their fears are warranted nine hundred and fifty thousand times per year, respectively.

Given safe bike lanes, more people—WAY more people—get on bikes and ride them more.[8] But it's complicated. There are few one-size-fits-all bike paths, given that there are so many reasons why people bike, where they bike, speeds at which they bike, and now with the advent of e-bikes—and electric skateboards, scooters, and hoverboards—what they consider to *be* a bike.

Around one hundred million people bike yearly in the United States; about fourteen million ride twice a week or more. The perfect bike path for someone commuting eight blocks in a large city looks different from that of someone riding twenty miles for exercise in the suburbs. But the one thing all safe bike lanes hold in common is they're best when "protected."

Protected biking lanes separate bikers from motorized vehicles by continuous barriers, planters, trees, broad boulevards, even lanes of parked cars. These provide a bounty of benefits for cyclists—most immediately a reduction in injuries, for cyclists, drivers, and pedestrians alike.[9] In New York City, crashes for road users of all

Bad: The painted bike lane offers no physical protection for the cyclist. Only the boldest cyclists are comfortable riding these lanes.

Better: The bollard-protected bike lane offers some degree of protection but few physical barriers from inattentive or turning drivers.

Best: The barrier-protected bike lane offers physical protection and ample space for cyclists, leading to increased ridership of young, old, and casual bikers.

SOURCE: WIKIMEDIA COMMONS

types—drivers, pedestrians, and cyclists—dropped 40 to 50 percent when protected bike lanes were installed, while roads with protected lanes generally see 90 percent fewer injuries than those without.[10]

A study in Portland, Oregon, divided the populace into four basic bicyclist categories. One percent were labeled "strong and fearless," 7 percent "enthused and confident," 60 percent "interested but concerned," and 32 percent "no way no how."[11] It's this middle 60 percent who, once their "concerns" are allayed, move toward "enthused and confident," and protected bike lanes are by far the best way to connect those dots. After Seville, Spain, constructed protected bike lanes, bike ridership increased elevenfold.[12]

There are other benefits. Rush hour is less rushed. In the Netherlands, 30 percent of the populace bikes to work, helping reduce road congestion; in the United States, that number is less than 2 percent.[13] The contrast is startling, but so is the fact that the Dutch have been building protected bike lanes for more than thirty years.

More younger and older people ride when there are protected lanes. In Denmark, where such lanes flourish, an elderly Dane is thirty times more likely to bike than his or her US counterpart. In Odense, Denmark, more than 80 percent of kids ride to school; the city's projected marker of success is creating protected bike lanes where six-year-olds feel safe riding to school alone.[14]

And businesses often thrive—even when parking is taken away to create bike lanes. In Salt Lake City, where nine blocks of downtown roads were revamped, resulting in the removal of 30 percent of the parking spaces, retail sales increased.[15] Studies show that while bikers may purchase less per shopping trip, they return to stores more frequently.

When protected bike lanes are in place, drivers feel more comfortable, roads are less congested, and fewer cyclists ride on sidewalks. Walker explains, "Good bike infrastructure makes cycling safer and more attractive, thus attracting more riders, in turn making cycling even safer. It is a virtuous circle.[16]

Also growing rapidly are multiuse or "shared" paths that accommodate bikers, joggers, wheelchair users, Segway drivers, in-line skaters, e-bikers, and walkers. The downside? Joggers with earbuds, dogs on leashes, tykes on trikes, and couples walking side-by-side often make such paths less-than-ideal places to bike, especially if one is commuting or on a mission.

Though most cost less than $100,000 per mile, protected bike lanes can be wickedly expensive; one recent downtown Seattle project cost $12 million per lane mile.[17] Protected bike lanes can also be wickedly complicated to build; some involve moving underground and overhead utilities, widening roads, installing new traffic lights, reconfiguring parking, even adjusting new building setbacks. And they can involve wickedly complicated long-term planning; many master bike path plans are on a twenty-five-year timetable, which can be a bit of a catch-22. Until you build paths, ridership won't increase, and until you build ridership, it's hard to justify more and better paths.

What do communities do when there isn't the space, funds, or appetite to build protected bike lanes? They paint them in.

Painted lanes are better than nothing, right? Well, yes and no. For starters, there are several kinds. *Sharrows* are standard roads, usually marked with chevrons and a bike symbol, where bikes and cars are intended to share the road equally. The mutual understanding is "Yes, bikes are slower than cars, but they get priority. No honking, passing, or acting like a jerk." *Striped paths* are dedicated bike lanes painted on roads that, as the name implies, separate bikes from traffic by a single stripe. *Buffered paths* feature an extra swath of paint and/or pavement and are intended to give bikers a greater physical and psychological distance from cars.

About 10 percent of cyclists—mostly those in the "young and fearless" and MAMIL (middle-aged men in Lycra) categories—feel "safe" riding between two painted lines. Commuter, recreational, young, beginning, and older cyclists don't. Because painted bike paths offer only a false sense of security, they don't build ridership.

The right-of-way at intersections—especially right turn lanes—is invariably confusing. One recent study showed that drivers passed fifteen inches closer to bikers in painted lanes compared with bikers on roads with no markings.[18] As Walker touts, "If mixing with motor traffic is your chosen bike environment, then almost all your cyclists will be a small group who are mainly young, predominantly male, and disproportionately gung ho."[19]

In a perfect world, every town would have a network of safe, sane, protected lanes. But perfect is not on the menu. On a recent bike jaunt through Minneapolis, I encountered a patchwork of bike lanes that were painted, protected with concrete barriers, separated with reflective posts, and sharrows.

Most problems (and accidents) arise in the gray areas. What's the best way to travel on roads less than fourteen feet wide where side-by-side car and biking becomes impossible? When is the best time to move over when making left turns? When is it okay for a cyclist to "take the road"?

Some will say, "If you're freaked out about biking, wear a helmet and bright yellow vest." Walker counters with the fact that helmets and high-visibility clothing "are a red herring, an irrelevance, a peripheral issue. You don't make cycling safe by obliging every rider to dress up as if for urban warfare or to work a shift at a nuclear power station. You do it by creating a road system that insulates them from fast-moving road traffic."[20]

Let's pretend we live in cities with protected bike lanes. What's in it for us? Walker imagines what life in his homeland of England would be like if he could press a magic button that would build protected bike lanes and increase the percentage of journeys taken by bike from its current 2 percent level to the Dutch level of 25 percent. For starters, he extrapolates, the increased daily activity could save fifteen thousand lives per year simply by making people less sedentary.

"But," Walker continues, "there's also reduced smog and the accompanying benefits in combating climate change, many fewer families destroyed by the grief of road deaths. . . . You can even factor in a notable boost to overall mental health, and more vibrant local economies." But best of all, Walker concludes, you'd create a people-based world rather than one "built for rapid, anonymous, one-ton metal boxes," where "people can amble, children can play, fresh air can be breathed, conversations can be heard."[21]

We need to get to the point where biking is an integral part of daily living—where people commute on everyday bikes, wearing everyday clothes, doing everyday things. On safe bike paths. We need to get to a mindset where the tubby old fart who passes us on his e-bike isn't a foe but a friend.

HACKS & FACTS ## Safety in Numbers

Statistics alone tell a large part of the biking safety story. Pay heed![22]

- Wearing a helmet reduces chances of head injuries by over 50 percent.

- Wearing high-visibility clothing reduces accidents by 47 percent.

- Alcohol is involved in 37 percent of bicycle fatalities.

- Male biking fatalities are eight times greater than female.

- Most bike fatalities occur between 6 p.m. and 9 p.m.

ASPHALT STREETS

Pavement, Potholes, *and* Mummy Paint

IN THE HISTORY OF THE WORLD, HAS THERE EVER BEEN ANYTHING more boring than asphalt? A line at the post office on December 20? A Kardashian tweet? Being stuck in traffic on the very substance we're about to explore? Inert, lackluster, ubiquitous—asphalt is the wallflower of wallflowers. Yet, we critically depend on it. When we walk around the block, there's a 95 percent chance the street we're strolling down is asphalt.

Asphalt, or bitumen, was the Flex Seal magic sealant of the ancient world. In its naturally occurring state, it was used to waterproof baskets, baths, reservoirs, and boats.[1] As an adhesive, it could secure stones together or weapon heads to handles. In ancient Egypt, it was used to embalm mummies. Eventually it became such a sought-after strategic material that the Seleucids and Nabataeans waged war over natural deposits of it in 300 BCE—the first of the many hydrocarbon wars, still popular today.

The melancholy stuff even had its artistic uses. It was used as a "photoresist" on early photographic plates. Painters found less success with it as an oil-based paint additive; it was discovered—albeit decades late—that it never fully dried and slowly and irreversibly pushed a number of the paintings of Eugène Delacroix, and those of other artists, in their march toward darkness.

In a bizarre jumble of asphalt, archeology, and art, "mummy brown" paint—ground-up, asphalt-preserved mummy mixed with white pitch and myrrh—was the rage in the mid-1800s; it had an unmatched aura and tone.[2] It was actually available up until the 1960s when a mummy shortage arose, though one of the last suppliers confessed, "We might have a few odd limbs lying around somewhere."[3]

Its dominance as a road paving material slowly evolved. Most early roads were little more than earth that had been flattened by horse-drawn plows plodding between surveyors' stakes. Early roads were filthy affairs.[4] Excrement from horses, pigs, and dogs blended with garbage and slop to create a roadbed of horrors. In summer, horses and wagon wheels pulverized the dried muck into a fine dust that infiltrated every home and every pore of occupants. In spring and rainy weather, streets devolved into a slippery mixture that wagons, horses, and pedestrians would slide across or sink into.

The cries of health officials and, surprisingly, the emergence of the bicycle craze in the 1890s prompted the movement for better roads. In many urban areas, the first pavement consisted of tree rounds set on end on a compacted base; the diamond-shaped space between would be filled with gravel. In other areas, four-by-four-inch blocks were installed, butcher-block style. Wood surfaces were quiet but wore unevenly, and in the presence of moisture, like rain, snow, and horse urine, hills became slippery affairs.

As wood roads wore out, they were often replaced by stone. Sandstone was popular since it provided good traction for horses, but it was soft and wore unevenly. Granite blocks were used in highly trafficked areas. They wore extremely well but were excessively noisy; some merchants complained they could barely hear their customers as wagons and horses clattered and clip-clopped by. On the East Coast, cobbles and squares of Belgian granite, originally used as ballast in ships' holds, were used; they were incredibly durable—indeed, some are still in use today.[5] But these created an uneven surface that was (again) slippery when wet.

In the 1910s, many cities that had converted to stone reverted to wood, this time creosoted blocks set on end, which proved both durable and quiet. One article in the July 1915 issue of *Sunset* magazine lamented, "A great part of man's happiness has been drowned in the rising sea of noise. Most of it attributed to hard pavement and cobblestones." But relief was on the way—"the wood-block pavement, silent as the hand of night, is coming into its own again."[6]

But could there be something better?

Natural asphalt, combined with burned brick, had been used as road paving in Babylon as early as 615 BCE. In the early 1800s, a few Scotsmen created *tarmacadam* pavement, using natural asphalt as the unifier. One of the first modern roads paved with asphalt was the Champs-Élysées in Paris in 1824. In the mid-1800s, entrepreneurs, inventors, and municipalities accelerated the development of asphalt, and by 1870, an asphalt tsunami was in motion. In 1876, Pennsylvania Avenue in Washington, D.C., was paved in asphalt for the national centennial.

Laura Ingalls Wilder, author of *Little House on the Prairie*, waxed poetic about asphalt in 1894:

> In the very midst of the city, the ground was covered
> by some dark stuff that silenced all the wheels and
> muffled the sound of hoofs. It was like tar, but Papa
> was sure it was not tar, and it was something like
> rubber, but it could not be rubber because rubber cost
> too much. We saw ladies all in silks and carrying
> ruffled parasols, walking with their escorts across the
> street. Their heels dented the street, and while we

watched, these dents slowly filled up and smoothed
themselves out. It was as if that stuff were alive.
It was like magic.[7]

Up until about 1900, most asphalt was derived from natural
sources like Lake Asphalt in Trinidad, which is still estimated to
contain ten million tons of asphalt. Asphalt can also ooze out of
the ground and be found in unlikely places like the La Brea Tar
Pits of Los Angeles. As demand exploded, asphalt processed from
crude oil became the norm. In the processing hierarchy, jet fuel is
at the top of the barrel, while asphalt is—literally—at the bottom.[8]
By the turn of the century the United States was crisscrossed by
thirty million square yards of asphalt paving. Asphalt technology
improved considerably during World War II as stronger, tougher
asphalts were developed to withstand the weight and impact of
military aircraft.

Today, thirty-five hundred mixing plants in the US churn out
about 350 million tons of asphalt paving material per year. If you
were to dig up every inch of asphalt paving in the country, you'd
have an eighteen-billion-ton pile around your ankles. The good
news is that all of it could be recycled—as 99 percent of it cur-
rently is—in the form of roadbeds, shoulder material, and new
pavement.[9]

Asphalt paving isn't pure asphalt—far from it. The surface you
walk and drive on is called asphalt concrete and consists of only 5
to 10 percent asphalt; the remainder is sand, stone, gravel, and
other aggregates. To create it, the aggregate is heated in a drum
tumbler to drive moisture out, then combined with heated asphalt.
Ground-up shingles, pulverized rubber, and recycled asphalt may
also be fed into the mix. The mixture must maintain a temperature
of 300 degrees F during transport, installation, and compaction—
thus you rarely see hot asphalt being installed in cold weather.

For some roads, a single layer of asphalt is installed over a cement or compacted base. More commonly, two layers, each two to three inches thick, are installed over a compacted subbase. After the first layer is installed, it's common to let cars in the neighborhood serve as involuntary steamrollers for several months as the asphalt and ground are compacted. Then a binder and a final top layer are installed.

As with so many other building processes—whether it be pouring concrete, laying sod, or installing brick pavers—ground preparation is everything. Before installing new asphalt, workers and loud machines run around for days ripping out old asphalt and putting in place new water pipes, sewer lines, and storm drains. New curbs may be built. Workers will make sure there's at least six inches of granular subbase, sloped so there's a 2 percent crown in the middle of the road for water runoff, then compact the bejesus out of it to minimize settling and dips. Only then will the asphalt be laid.

After it has cooled and been compressed by a ten-ton road roller, asphalt can be driven over almost immediately. A well-built asphalt road will last fifteen years or more.[10] Although a concrete roadway lasts longer, an asphalt road costs a quarter to half as much as concrete and is more easily cut into and repaired when accessing the water, sewer, and gas pipes underneath it.[11] Scientists and engineers are working on self-healing, snow-melting, and noise-reducing asphalts.[12] The increasingly popular "perpetual asphalt," which is composed of three engineered layers of asphalt—each with its own particular strength—has a projected life of fifty years.

It may lack glamour but it's a miracle product—almost.

As asphalt dries and flexes under vehicle weight, small fissures occur. Water infiltrates these cracks and softens the underlying road base. As the moisture freezes and expands, it forces the asphalt into small chunks, which are pushed up and out by vehicle traffic and frost heave.

The American Automobile Association maintains that around $3 billion of pothole damage is inflicted on three million vehicles annually in the United States. This damage manifests itself in bent wheels, damaged suspension, exploded tires, and mangled vehicle and human bodies.[13] Bicyclists, pedestrians, Segway operators, and even motorized wheelchair owners have wound up in the ER or, less frequently, the morgue. It's a conundrum: pothole repairs are best done in warm weather, but potholes most likely form when it's cold.

Pothole horror stories abound. One Tesla owner hit a pothole and blew out the entire suspension system. As her car was being towed, another car hit the same pothole and was rear-ended by yet another car. New York City publishes the online "Daily Pothole," which allows you to check the condition of your street and report potholes, hummocks (lumps), cave-ins, and missing manhole covers. The Maryland Department of Transportation has a pothole hotline and tries to fill reported potholes within forty-eight hours, but with more than one hundred thousand potholes to fill each year, the task is Sisyphean.[14] Plus, the cold asphalt patches made during early spring and winter are notoriously short-lived. Potholes so devastated the Baltimore-Washington Parkway in 2019 that even after filling potholes with sixty tons of asphalt, the US Park Service had to lower the speed limit from fifty-five to forty miles per hour.[15]

You can seek damages from whatever governmental agency has jurisdiction over the offending road, but it's a bumpy battle. Toronto, for example, denies 96 percent of claims.[16] Potholes must have been previously reported and gone unrepaired for an unreasonable length of time for a claim to be considered.[17]

The problem is epidemic worldwide. In the United Kingdom a car is damaged by a pothole every eleven minutes.[18] In a recent five-year period in India, 14,936 deaths were attributed to potholes, or around 10 per day.[19] A combination of poorly maintained roads, monsoons, a preponderance of two-wheeled vehicles, and no laws to hold road contractors or municipalities responsible has created a calamity that continues to grow worse. One father—who lost his sixteen-year-old son when the motorbike he was on hit a water-filled pothole—has taken it upon himself to repair more than five hundred potholes. "Our nation has a huge population," Dadarao Bilhore wrote. "If even one lahk [one hundred thousand] of us started filling potholes, India will become pothole free."[20]

HACKS & FACTS Click for a Quick Fix

See a pothole that needs filling? If you live in a community that's enrolled in SeeClickFix, you can report it with one click of your mouse or tap on your tablet. The SeeClickFix website (https://www.seeclickfix.com) directs your concerns to the right government official pronto.

Private citizens and companies have taken matters into their own hands. Domino's Pizza initiated its "paving for pizza" pothole-repair program ("bad roads shouldn't happen to good pizza") and offered $5,000 grants to cities in all fifty states for repairs.[21] A person in the Netherlands has taken to planting flowers in potholes to draw attention to them. And Frank Sereno of Kansas City recently held a birthday party, complete with birthday cake and candle, to call attention to a long-unfilled pothole by his house. Local news media covered the party. The hole was rapidly repaired.

ALLEYS

One Man's Love Affair

PITY THE POOR ALLEY—NAMELESS, NEGLECTED, REPOSITORY OF all things nasty; a shadowy place where rats scurry, muggers lurk, and only the graffiti has color. It's not a place we want to get to—just through. But not so for Christian Huelsman. He has a love affair with alleys. He wants you to too.

"Alleys can be amazing assets to a community," Huelsman enthuses. "If you clean them up, give them names, add lighting, include some public art, give them a sense of place, well, people will use them." Huelsman, an alley advocate, should know; he's done just that. As founder and executive director of Spring in Our Steps, Huelsman and crews of volunteers, armed with rakes, shovels, and brooms, have spent years cleaning up forty years' worth of construction debris, junkie needles, and broken bottles from alleys and public stairways across Cincinnati. "Some alleys were so full of crap we had to rent a Bobcat," he tells me. His group then began hosting alley-based community events: film screenings, photo exhibitions, and musical performances. "Alleys and buildings have historical attributes and a grittiness that fronts of buildings have often lost through renovation. If you make alleys accessible, people love to explore them."

Huelsman goes on to explain how, once the stage is set, food trucks and pop-up shops can be included. Smaller scale buildings, artist spaces, and much-needed accessory dwelling units could be built to add diversity and vibrancy to a block and community. Four communities have done just that.

San Francisco's Chinatown. The alleyways of Chinatown—historically crammed with butcher shops, cigar factories, pawnshops, and tenement entryways—have always been alive. But the Chinatown Alleyways Renovation Program is making them even more vibrant with public seating, landscaping, murals, and pop-up shops. Some are now open only to foot traffic. And some are named, themed, and have their own unique personalities—like Jack Kerouac Alley, named after the famed beat writer, and Salt Fish Alley, known for its fish shops.

Seattle's Alley Network Project. Public and private sectors have joined to invigorate the alleys around Pioneer Square in downtown Seattle. Retail stores, restaurants, poetry slams, and World Cup soccer screenings have become part of the scene. Nord Alley has been crowned a Festival Street and is closed to traffic on select days for celebrations; Pioneer Passage has developed into one of the city's greatest pedestrian spaces.

Chicago's Green Alley Movement. The Windy City, with nineteen hundred miles of alleyways—more than any city in the world—has plenty of square footage to work with, and plenty of rainwater to dispose of. Step one has been to repave alleys with permeable concrete and asphalt made of recycled material so rainwater seeps into underground water tables rather than becoming oily runoff. Using lighter colored paving material, heat is reflected, helping cool the city; new energy-efficient "dark sky" lighting eliminates light pollution.

Melbourne's laneways. The economic downturn of the 1990s forced Australia's second largest city to rethink its underused spaces—laneways in particular. The opening of Meyers Place on a small laneway led the charge with cafés, restaurants, and quaint stores. Laneways have become "the beating heart of Melbourne," crows Karl Quinn of the *Sydney Morning Herald.* "Who could fail to see the romance inherent in a cobbled street built for carting away crap by night, right?"[1]

San Luis Obispo's Bubblegum Alley has attracted two million wads of gum since its inception twenty years ago. It also attracts one hundred thousand visitors per year, making it one of the most frequented tourist spots in the city (and proving alleys can be something special).

Missing from the vibrant-alley list is New York City, for one good reason—the city doesn't have any alleys (well, one resident counted three). When the original street grid was laid out, planners determined that land was too valuable and the population too high to waste valuable space. The tradeoff has been piles of garbage lining the sidewalks at night and deliveries having to be made via basement elevators popping up through sidewalks.

There's something inherently quirky about alleys. We don't expect much from them, so when they sparkle even a little, we take notice. An alley can help give a neighborhood or city an identity.

Formerly neglected alleys like this were spruced up to become the lifeblood of many restaurants and cities during and after the COVID-19 pandemic.

People become interested in alleys when they're presented right. Huelsman conducts sell-out bicycle tours of alley street art. Transforming an alley takes work and vision. From the start they've been more utilitarian than utopian: places for coal trucks to unload, servants to enter, horses and carriages to be housed, and trash bins to be stowed. Rethinking these spaces is a worthwhile exercise. "Alleys can be full of surprises," Huelsman says. "We should embrace measures that will make these surprises ones that activate, captivate and inspire."[2]

CONCRETE

Sidewalks, Dams, *and* that Damn Joan Crawford

WALK AROUND HOLLYWOOD AND EVENTUALLY YOU'LL FIND YOUR-self standing on the sidewalk in front of the famed Grauman's Chinese Theatre. Each concrete slab holds its own impressive story. Comedian Mel Brooks donned a prosthetic sixth finger before making his handprint. Marilyn Monroe dotted her *i* with a rhinestone. Joan Crawford inscribed, "May this cement our friendship"—though masons 'round the world cringed at this common muddling of the terms "concrete" and "cement."

A stretch of sidewalk between our house and Nelson's Ice Cream three blocks away tells another interesting, albeit less starstruck, story—that of a curious dog. You can follow the paw prints as he or she pauses upon entering the wet concrete, then traipses over to a tree before zigzagging through another twenty feet of wet concrete and bounding off to the side. There's the "curious dog" part of the story, but also the "incredible material" story. Here's a substance that in the morning was a pile of rocks and powder, by coffee break was mushy as peanut butter, by lunchtime was "plastic" enough for a canine to stroll across it, and by the next day was hard enough for humans to walk over—for the next twenty-five years. Concrete is a material that's precious, yet cheap; brawny enough to build the Three Gorges Dam on the Yangtze River, yet friendly enough for you to mix up a batch to set a fencepost.

About ten billion tons of concrete—one-and-a-third tons for each man, woman, and child on the planet—are produced every year, making it second only to water in terms of how much is "consumed."[1] All in all, concrete constitutes half of everything we build.[2] Its usage, ton for ton, is double that of steel, wood, plastic, and aluminum combined.[3] Your one-and-a-third tons aren't delivered to your doorstep by UPS, but you use it just the same. You walk and drive over and under it, you work and live in buildings made of it. It's used for sewers, countertops, boats, skyscrapers, and the sidewalks upon which you—and your inquisitive dog—take your daily walk.

But what the hell is it? And who's responsible for misplacing the recipe for nine hundred years?

To placate masons and lexicographers around the world, we start by clarifying the difference between cement and concrete. *Cement* is a pasty mixture—think melted marshmallows—consisting primarily of water, lime, silica, iron, and alumina. *Concrete* is a chunky mixture—think Rice Krispies Treats—created by adding sand, stone, and/or aggregate to cement.

Cement wears many hats; the *mortar* used for laying bricks consists of cement mixed with sand; the *stucco* and *plaster* used on houses is cement mixed with different proportions of sand and lime (and, in the past, horsehair); even the *thinset adhesive* used for setting tile contains cement. Though there have been many, many formulas for cement through the ages, they ALL contain one essential ingredient: lime, or calcium oxide. Lime is the defining characteristic—the sine qua non—of cement. It's what holds everything together—the melted marshmallow that bonds the Rice Krispies Treats together. Indeed, the word "lime" comes from the Old English word meaning "sticky substance." No lime, no cement.

But one doesn't just find lime; with rare exceptions, cement is created by heating limestone or other calcium-rich substances to 1,600 degrees F, then grinding it into a fine powder. The world has plenty of limestone; it constitutes 4 percent of the earth's crust and is found in places as unlikely as the summit of Mount Everest. It's made of millions of generations of shellfish and coral that have been crushed, compacted, and crystallized over hundreds of millions of years. The trick is creating that 1,600 degree F temperature.

Twelve-million-year-old natural concrete and ten-thousand-year-old crude manmade versions have been discovered. Some people suspect that the earliest forms of lime were created by lightning strikes, volcanoes, or roaring bonfires perched beneath limestone outcroppings. Six thousand years ago, lime mortar was used to cover the outer surfaces of Egyptian pyramids. A few will argue—though not successfully—that the pyramids themselves were built using cast-in-place concrete.[4]

But most credit the Romans with perfecting—well, almost perfecting—the art and science of the substance from 300 BCE to 475 CE. They were successful because they developed wood-fired kilns in which they could precisely control the temperature and duration of limestone firing. They also fiddled with the proportion of water and other added materials. Initially, they mixed in sand and built only floors and short walls. When Romans started adding larger stones and aggregates, they started having a field day with the stuff. They even figured out how to create hydraulic cement that would set up under water.

Some people will argue that Roman concrete, because it incorporated *pulvis puteoli* volcanic powder, was actually superior to today's concrete. But two things prevented Romans from using concrete to its full potential. First, they lacked a way to thoroughly mix all the ingredients, as today's massive concrete trucks do, to create a strong homogenous substance; therefore, they built structures by layering cement and aggregate and then compacting the devil out of the materials as they were added. Second, they didn't reinforce their concrete with metal bars (rebar) or mesh as we do today—not surprising, given the status of metallurgy back then. As a result of these two deficits, they had cement that was excellent in compression (in the form of pillars, domes, and arches) but lousy in tension (in the form of horizontal beams). It's similar to wood in this respect. The skinniest wood chair legs can support a five-hundred-pound person all day long because wood is at its strongest when it is oriented vertically like a pillar in compression. Take that same leg, suspend it horizontally like a beam by the ends, and have that same five-hundred-pounder sit on it, and it's "tweezers please."

That didn't stop the Romans from building structural and aesthetic masterpieces. The Pantheon, built around 100 CE, continues to hold the record as having the world's largest unreinforced concrete dome. The thirty-foot-diameter open oculus at the top and the 140 coffered panels adorning the inside make the remarkable even more remarkable. Upon seeing the massive span, many architects and engineers mistake it as a modern reproduction.[5]

The Roman Colosseum, Pont du Gard aqueduct, and Nero's Golden House are other examples of ancient Roman structures made at least partially with concrete that are still standing, albeit vastly restored, today.

But a funny thing happened on the way to the forum: when the Roman Empire went on the skids, so did the use of concrete. People became too busy fighting off starvation, plagues, political upheaval, and one another to focus on architecture. It was a dark age indeed. Books were destroyed, including Vitruvius's *On Architecture*, one of

The 142-foot-diameter Pantheon has held the record as the world's largest unreinforced concrete dome for twenty-one-hundred years. The dome tapers from twenty-one feet in thickness at the bottom to four feet at the top.

the few books containing the formula for concrete. For a thousand years, most structures were constructed out of wood or stone. Some mortar-like materials were used, but magnificent concrete structures the likes of those built during the golden age of the Roman Empire ceased to be.

In 1414, a copy of *On Architecture* was discovered in a Swiss monastery. This discovery, along with the dawning of the Renaissance, got concrete flowing again. Concrete bridges and buildings began appearing. Still, improvements were slow in coming.

In 1824, an English bricklayer from Leeds named Joseph Aspdin was granted a patent for Portland cement, which involved a new method of combining and heating the requisite clay and limestone unearthed from the rich grounds of Portland, England. (That city in Oregon has nothing to do with the name.) Aspdin acquired the then-expensive limestone needed for experimenting by stealing pavers from the streets of West Yorkshire.[6] Today, most cement is one of five Portland types available.

Thomas Edison helped increase the visibility of Portland cement by developing and promoting a line of concrete houses. He also tried his hand at concrete furniture, which he claimed would be more durable and artistic than oak and cost half as much. He shipped a prototype concrete phonograph cabinet on a round-trip journey through Chicago and New Orleans with a shipping crate label that read, "Please drop and abuse this package." The planned unveiling of this well-traveled cabinet at a press conference never occurred, presumably because the cabinet was indeed dropped and abused.

Another genius, the architect Frank Lloyd Wright, had considerably better luck. When an earthquake shook Tokyo on the opening day of Wright's concrete Imperial Hotel, the hotel remained relatively unscathed, while most wood structures around it collapsed and burned.[7]

The concrete in your driveway, sidewalk, or road outside your door today contains 10 to 15 percent cement, 60 to 75 percent aggregate and sand, and 15 to 20 percent water. It may also contain fly ash (a byproduct of coal-burning power plants), slag (a byproduct of steel manufacturing), air, and other additives.[8] Concrete doesn't become solid by "drying out" but rather through hydration—a chemical reaction in which the powdered cement and water combine to crystallize and harden, locking the sand and aggregate in place.

The proportions and purity of the ingredients are critical. Increasing the ratio of cement to water creates a stronger concrete—to a point. If the mix is too dry to coat and bind the aggregate, the concrete is too weak. Conversely, if you add too much water or too much dirty water, the concrete is too weak because the cement crystals are too far apart in the mix to bond the sand and aggregate together—a deadly problem as we'll soon discover.

Ninety percent of concrete's strength is reached during the first few weeks; the remaining 10 percent can be achieved over years or even decades. Some of the four million cubic yards of concrete used to build the Hoover Dam in the 1930s is still curing.

Today, engineers, architects, and contractors install *rebar*—rods of metal with ridges—in the concrete to absorb the tension created by weight, gravity, shrinkage, and movement. If you visit a sports stadium, dam, bridge, or other larger scale project under construction, you'll encounter a jungle of rebar. Contractors may also install panels of welded wire mesh to minimize cracking in slabs.

Building with concrete involves taking natural rock, heating it, grinding it, and mixing it with water and more rock to create a viscous form that can be placed anywhere, in any shape you please—before it hardens back into rock again. Need a three-thousand-mile freeway capable of supporting fifty-ton trucks for thirty years? Use concrete. Need an opera house in Sydney that'll make people applaud even before they step inside? Concrete. Need a slab for your charcoal grill, a curvy countertop for your kitchen, or steps outside your front door? Start mixing.

You cannot avoid concrete in your day-to-day life. The concrete floors and sidewalks you stroll across are normally four inches thick, the driveway you drive down is six inches thick, and the interstate you speed along is eleven inches thick. To give conspicuous slabs of concrete an inconspicuous space to crack as it expands and contracts, control joints—one-inch or deeper grooves—are spaced every few feet.

Just as Rice Krispies Treats will be gooey, nummy, or rock hard based on how well the recipe has been followed, concrete will be strong or weak depending on the proportion and quality of ingredients used.

Examples of poorly mixed concrete are legion—and disastrous. Many of the estimated two hundred thousand people killed in the 7.0-magnitude earthquake in Haiti in 2010 were crushed by poorly built masonry structures. The primary problem was weak concrete, the result of contractors using too much water or too little cement. "Most buildings are like a house of cards," Roger Musson of the British Geological Survey explains. "They can stand up to the forces of gravity, but if you have a sideways movement it all comes tumbling down."[9]

Another problem has to do with rebar. In many ways, concrete and rebar are a perfect duo because they expand and contract at roughly the same rate, and rebar, encased in concrete, would seemingly be protected. But nature is a formidable adversary. Freeze-thaw cycles, saltwater, road salt, movement, and time can create small fissures in concrete that invite water to seep through, rusting the rebar. When rust expands, it compounds the cracking. Because rebar is buried in concrete, it's difficult to determine its state. Contributing to the problem, particularly in developing countries, is the lack of building codes and enforcement. This Achilles' heel of concrete doesn't negatively affect newer, well-designed structures much, but older bridges, piers, roads, power plants, and other massive structures are at risk. Estimates for repairing or replacing the degrading concrete infrastructure of the United States alone is in the trillions of dollars.

Concrete has an environmental impact that's hard to see, but very real. Its formulation is responsible for 4 to 8 percent of the world's carbon dioxide emissions. It uses about 10 percent of the world's industrial water supply, straining the amount of water available for drinking and irrigation. In cities, it contributes to the heat-island effect by absorbing and radiating heat from the sun. And the dust generated during its mining, manufacture, and mixing—especially in developing countries—contributes to respiratory disease.[10]

While there are few substitutes for concrete, developers are trying to make concrete last longer, mitigating the need to create more of it. A new form of "self-healing" concrete is under development. This process involves adding bacteria, which react with water to excrete calcite that in turn bonds to concrete. This helps seal cracks. Developers are also learning that adding titanium dioxide to concrete creates a self-cleaning surface and, remarkably, one that can remove smog, nitrogen oxides, and other pollutants from the air.[11]

Corrosion-resistant rebar made of aluminum bronze is projected to last several hundred years. The initial cost is about 20 percent higher than that of traditional rebar, but compared with replacing a bridge every seventy-five to one hundred years, it would save hundreds of millions of dollars. One study shows that the repeated cost of replacing a typical bridge made with standard rebar over five hundred years might be more than $300 million, while the "lifetime cost" of the bridge made with improved rebar would be closer to $70 million.[12]

The concrete jackhammered from old roadways and structures is increasingly being crushed and reused as aggregate in new concrete or as base material under the concrete. Recycling concrete saves landfill space, reduces the need for mining, reduces transportation costs, and can reduce the cost of aggregate by half.[13] More than 140 million tons of concrete are recycled annually in the US alone.

There's little chance a substitute for concrete will be created in the near future, but there's a great chance we can make our structures stronger, longer lived, and more environmentally friendly.

And I'm quite certain Joan Crawford would approve of this.

HACKS
& FACTS

A Concrete Way
to Increase Gas Mileage

Vehicles traveling fifty miles per hour or faster get 8 percent better gas mileage on roads made of concrete than on those made of asphalt.[14]

PARKING

The Secret Cost
of Free Parking

I'VE NEVER MET A PARKING METER I LIKED; NEVER SLID A QUARTER in the slot and had three cherries line up. Same holds true for lots and ramps; never swiped my credit card and gotten a warm hug.

That's why I love FREE parking. I love it because 97 percent of parking in the United States is FREE. I love it because when my vehicle, like most, is parked, 95 percent of the time it's parked for FREE! FREE! FREE!

Or is it?

Let's see what the numbers, history, and the "prophet of parking" have to say.

Surely, the meters and pay stations lining our city streets aren't FREE, FREE, FREE. All totaled, parking meters in this country gobble down more than $30 billion in revenue a year.[1] We have Oklahoma City to thank for that.

The downtown area was having a problem: Employees of local businesses were arriving early, claiming all the nearby parking spaces, then hogging them all day. Customers had to park blocks away and then haul their wares long distances. In 1935, newspaperman Karl

Magee presented a solution: place his spring-powered, 5-cents-an-hour Park-O-Meters at twenty-foot intervals along the streets.

Retailers were wary. Would meters drive customers away? No. In fact, the resulting quicker turnover of cars and customers increased business appreciably. By the early 1940s, more than 140,000 parking meters were ticking away throughout the United States.

With its red pointer and EXPIRED flag, the basic coin-operated meter changed little over the next forty years. Meter collectors gathered the money from those who plugged the meters, and meter maids, armed with parking ticket books, collected money from those who hadn't.

Life—or at least parking—seemed simple.

During the 1990s, electronic, digital, and multispace meters clanged onto the scene. They could be fed with credit cards, debit cards, and mobile apps. Some featured smart cameras that photographed license plates and issued tickets for expired meters. Over time, wireless systems that communicate with your vehicle started charging only for the minutes drivers used. Today, plugging the five million meters in the US is increasingly becoming "frictionless." We no longer need pockets or ashtrays bulging with quarters. "In the future," parking guru Donald Shoup tells me, "the way we handle parking today will seem Victorian."

Parking meters are a tidy source of income for many cities. Denver collects about $12 million a year in parking meter income; even little old Charleston, South Carolina, bags $4 million.

Cities also make plenty of coin from tickets issued for expired meters and illegal parking. The average American gets one parking ticket every five years; Los Angelenos get five times as many.[2] In recent years, the city of New York issued nearly twelve million parking tickets, pulling in $534 million in revenue, the highest amount of any US city.[3] One industrious meter—the TM416 in

Amazing Grace

Most parking meters have a two- to ten-minute grace period built in before expiring. Also, if you contest a parking ticket—and it's your FIRST one—there's a 25 to 50 percent chance the judge will waive the fine.

front of the El Paso County Courthouse—is an overachiever: in two years, the meter was responsible for the issuance of 293 parking tickets.[4] Not all parking violations are created equal. In San Francisco, 43 percent of all parking tickets are the result of "street cleaning." Meanwhile, in Boston, nonresidents parking in "residents only" areas garner the most tickets.

Parking is expensive in other ways. A study by INRIX, a company providing parking information to smart cars, estimates that each driver wastes about fifty-five hours a year looking for parking spots, which results in $600 million in wasted fuel.[5] One composite of twenty-two studies showed that between 8 percent and 74 percent of the cars in congested areas were cruising and circling for a parking space—and it took anywhere from three-and-a-half to fourteen minutes to nab one. Another negative side effect of parking space cruising is drivers are notoriously negligent of the pedestrians and cyclists around them. And angst? More than 60 percent of American drivers report avoiding certain destinations because of a lack of available parking. An equal percentage report that searching for a spot causes increased levels of stress, while almost one-quarter of drivers have experienced "parking rage." And one study estimates that Americans spend 3.14 billion hours a year bitching about parking.[6]

But by far, our biggest expense for parking is the creation of spaces themselves. We build them like ants on amphetamine. One study cites there are eight parking spots for every car in the United States. In Houston, there are thirty parking spots per resident, which doesn't help a whit in the event of disasters like Hurricane Harvey, which dumped nine trillion gallons of water on the city in two days, much of it with no place to go.[7] In the US, there are 1,300 square feet of off-street parking per car, but only 720 square feet of housing per person.

The cost of building each parking space starts at about $20,000 for paved-surface parking and soars to $75,000 and more for underground spaces. One way or another, those costs are passed on to the consumer or resident—even if that consumer or resident doesn't own a car. What to do?

Ask Donald Shoup—the "prophet of parking," professor of urban planning at UCLA, and author of *The High Cost of Free Parking*. He suggests that we look not only at the cost of the parking we pay for, but also at the cost of the parking that's "free"—that is, the financial, societal, and environmental costs. Shoup maintains that the dollar amount we pay for "free" parking in the United States falls somewhere between what we pay for Medicare and national defense.[8] To make cities less congested and make parking less of a hassle, Shoup proposes three parking reforms, all of which create a healthier economy and environment.[9]

First, Shoup says parking meters should be governed by the market economy. When demand is high, meter and parking lot prices should be high; when demand is low, prices should follow. Many US and European cities have already rolled out variable pricing. "Peak rates" during times of high demand reach as much as $18 per hour; in moments of lull, a person might pay $0. The overall goal is to price meters so one or two open parking spaces—

about 15 percent—are available on any given street at any given time. It can be complicated. Shoup reminds us there's more and more "competition for the curb" these days; curbs need to accommodate not just parking, but bus stops and bike lanes, ride-share and taxi pickup sites, mini-parks, loading zones, and the ever-increasing number of delivery trucks.

The second step is to reinvest the parking revenue back into the area where the meters are located so people can see their dollars making a difference. Projects can include improving sidewalks and streets, building bike lanes, improving public transportation and education, installing free Wi-Fi, and providing benches, trees, and green areas for people to enjoy.

The third step is to remove off-street parking requirements mandated by cities—requirements that create too many parking spaces, make cities less walkable, and indirectly increase the cost of services and goods. Almost every city's municipal code has a "Minimum Parking Spaces Required" section. Looking at section 21.22.120 of the municipal code for Monterey Park, California, we find that office buildings are required to provide five-and-a-half parking spaces per thousand square feet of floor space; restaurants, ten; and tattoo parlors and fortune-telling establishments, four.[10] Golf courses need five parking spaces per hole, while hospitals need three per bed. These standards may not seem excessive, but each parking spot on average requires 330 square feet for the car and common spaces for entering and maneuvering.[11] In other words, for every thousand square feet of restaurant, thirty-three hundred square feet of parking is required, or three-and-a-third times more space for parking than gobbling. Parking requirements make your hamburger, movie ticket, and apartment just that much more expensive.

Minimum parking requirements also contribute to urban and suburban sprawl by creating buildings sitting in the middle of vast parking lot wastelands. In retail settings, determining parking lot size based on one day of the year—the Saturday before Christmas—is disingenuous. Minimum parking requirements also hamstring developers wishing to convert older historic structures into living spaces in downtown areas, where land is either too expensive or too sparse to provide the required amounts of off-street parking.[12] Fewer and smaller parking lots would put buildings closer together and make cities more walkable. Urban planner Jane Jacobs wrote back in 1962, "The more downtown is broken up and interspersed with parking lots and garages, the duller and deader it becomes, and there is nothing more repellent than a dead downtown."[13]

On the bright side, many cities are changing minimum parking requirements.

As people, meters, parking lots, and cars become more and more connected, the face of cities and parking will change. How? Meters and lots could include cameras and sensors to detect occupancy, information that can be relayed to central computers that would adjust meter and lot rates up or down on the basis of demand. Those same computers could send information to smart phones and smart cars (and smart drivers) to inform them where openings were and how much parking would cost and to suggest alternative ways to get there.

Parking fees might pay for new bike lanes, walking paths, trees, and public art—resulting in less congestion, fewer accidents, and lower CO_2 emissions. Maybe then I'd meet a parking meter I actually liked.

HACKS & FACTS ## Passive-Aggressive Parking

A study shows that drivers take 21 percent longer to leave a parking space when someone is waiting for it and 33 percent longer if the waiting driver honks.[14]

WALKING

Soles for the Body,
Mind, *and* Soul

PICK THE HAPPIEST MAN YOU KNOW, JACK HIM UP TO SIX-FOOT-three, then slap on a day-glow shirt and a Sam Elliott moustache—and you'll start getting a picture of Dan Burden. I'm standing on a sidewalk in Salinas, California, with Burden—a man with the awesome title of "Director of Innovation and Inspiration/Blue Zones" on his business card—participating in a walking audit. It's my first audit, Burden's 6,382nd. We're not alone. Thirty other organizers, teachers, health-care managers, city employees, and members of grassroots movements are shuffling and striding too.

The walking audit is part of a daylong planning "charrette" at Salinas City Hall, where Burden and a half dozen other members of the Blue Zones team are leading discussions about steps they can take to make Salinas a better place to, well, take steps in—and live in, work in, and bike around.

We near an intersection. Burden pauses, raises his pointer finger, and asks, "Why do all clocks move clockwise?" Silence. "Why are electric outlets fourteen inches off the floor?" Silence. "Why are all train tracks spaced the distance they are?" Silence. Burden explains that clock rotation is based on the movement of sundials,

Dan Burden measures a crosswalk and explains, "We need comfortable places for people to walk. There are lots of better ways to protect pedestrians than with white paint."

PHOTO COURTESY OF BLUE ZONES

outlet height on the length of the hammer handles used by the electricians who install them, and train track spacing on the width of two horses' asses—horses that pulled chariots twenty-five-hundred years ago. What do sundials, hammers, and horses' asses have to do with walking?

"We get used to doing things traditional ways and keep on doing them even when they no longer make sense," Burden says. Rather than making our roads faster, wider, and vehicle-centric, Burden thinks—no, KNOWS—we should make roads and sidewalks more people-centric.

He breaks out a tape measure. The width of the sidewalk is forty-eight inches—because that's tradition. "But if we're walking side by side or I pass someone or there's a kid on a bike or there's a tree planted in the middle, we need five feet—better yet, six feet—to comfortably pass," Burden explains. He waves his hand at the sterile four-lane road abutting the curb and says, "We need less of that," then, tapping his foot on the sidewalk, "and more of this."

He points out pedestrians scurrying across an intersection to navigate sixty feet of roadway to safely cross to the other side. He then points out another intersection where peninsulas bulb out into the street to make the crossing a more pedestrian-friendly forty feet. "We need comfortable places for people to walk," he says. "And it's not just the sidewalk; it's having enough cool design elements on the block that you *want* to walk there. You need trees, things to see, places to sit."

A proponent of "street diets" that make roads narrower rather than wider, Burden isn't railing against the internal combustion engine; he's railing for good planning. Converting four-lane city streets to two-lane boulevards, and then using the newly created space to build bike lanes and walking paths, adding trees and replacing stop signs and lights with roundabouts, can make cities safer, flow better, friendlier, less congested, and more "people scaled."

"Walking is the core to everything about designing a community," Burden explains. He talks about active transportation. What happens when we make it easier and safer for people to bike, hike, and use public transportation? There's increased physical activity, more social interaction, increased property values, less pollution—and the list goes on.

The day before the charrette, a second-grader walking to nearby Sherwood Elementary School was struck by a car and seriously injured. If you're a kid—or any pedestrian—and get hit by a vehicle traveling twenty miles per hour, you have a 90 percent chance of survival; but at forty miles per hour, you have a 90 percent chance of being killed. In 1969, nearly 90 percent of children living within a mile of school walked or biked there; today, that percentage has plummeted to 30 percent.[1] After twenty years of declining pedestrian deaths, the fatality rate has soared over the past decade, climbing from 4,300 in 2009 to 6,590 in 2019. Likely factors include cell-phone use by distracted walkers and drivers, headphone "deafness," increased numbers of people walking for exercise, and alcohol. In 2018, 33 percent of pedestrians killed in vehicular accidents had a blood-alcohol level exceeding 0.08.[2]

Solutions are discussed. One participant talks about a city where orange flags are stashed in bins at stoplights so people can wave them to increase their visibility in crosswalks. Burden politely bristles at the thought. "People shouldn't have to wave flags or wear day-glow vests or run to get safely across a street," he says. "We can do better than that." Simple solutions include better lighting (three-quarters of pedestrian accidents happen at night), raising crosswalks six to twelve inches (so drivers can see pedestrians better and encounter a physical object to slow them down), and programming stoplights so pedestrians have a four- or five-second head start on cars to minimize "turning" accidents.[3]

The idea of the "walking school bus"—where adults accompany students on their walk to school, with the number of "bus occupants" growing in number and visibility block by block—is discussed. "Bicycle trains" work too. Safer routes equals more kids walking, and walking equals healthier kids, which means kids who become walking adults, adding up to healthier communities.

Walkable cities don't happen overnight or by accident. Burden and team discuss the "life radius" approach to community-building—the concept that when parks, schools, stores, friends, libraries, and coffee shops are within a one- to two-mile radius of where we live, we're more likely to walk or bike. Burden explains that fewer than 4 percent of people nationwide commute by walking or biking; but studies show that more than half the people would walk or bike if safe trails were provided.

Burden and his "built-environment" team are one branch of the larger Blue Zones initiative. The other two major branches are healthier eating and less tobacco use. The Blue Zones concept was born when Dan Buettner—after setting world records for distance biking, including a 15,500-mile trek from Prudhoe Bay, Alaska, to Tierra del Fuego, Argentina—set out to discover the places where people lived the longest. He found five communities— Okinawa (Japan), Icaria (Greece), Sardinia (Italy), Nicoya (Costa Rica), and the Seventh-Day Adventists of Loma Linda, California—that were longevity "hotspots."

In partnership with National Geographic, Buettner and team studied the cultures, interviewed people, lived in the communities, and looked for commonalities to longevity. Nine themes emerged, including being strongly engaged in family, social, and spiritual life; focusing on a plant-based diet; drinking a glass of red wine daily; and reducing stress. At the top of the heap sat daily physical activity. Aislinn Leonard of the Blue Zones explains: "The world's longest-lived people don't pump iron, run marathons, or join gyms. Instead, they live in environments that constantly nudge them into moving without thinking about it. They grow gardens and don't have mechanical conveniences for house and yard work. They have jobs that require them to move or get up frequently. And they walk every single day." And when we live in cities and have lifestyles that don't require weeding gardens, pinning clothes on a line, hauling in fish nets, and other daily movement, walking can be the key to providing that vital daily physical activity.

By some accounts, walking is the No. 1 form of exercise in the United States—though this statistic is akin to the view that lawn mowing is the No. 1 "hobby." Like breathing, walking is an integral part of living. It's the first activity toddlers want to master and the last thing the elderly want to give up. We all use it to trek short distances. But what about going that extra mile? What's in it for us?

Health benefits. More studies have been done on the benefits of walking than any other form of exercise. The upside is indisputable. Benefits are linked to three basic factors—intensity, duration, and frequency—but age, weight, terrain, stride length, overall fitness, weather, footwear, motivation, elevation, even socks, all affect the results. Nonetheless, walking can be clumped into four broad modes.

1. ***Restorative mode.*** If you are recovering from an injury or illness, are chronically sedentary or overweight, or are just trying to reach baseline health, the benefits of walking are enormous. Doctors often prescribe walking because it requires no special equipment, training, or facility. Progress is easily measured by minutes or miles increased, and you reap the important side benefits of getting some sunshine and socializing. Robert Sallis, a sports doctor with Kaiser Permanente, states that walking is "the best thing we can do to improve our overall health and increase our longevity and functional years."[4]

2. ***Maintenance mode.*** If you're in reasonably good shape, walking at a two- to three-miles-an-hour clip (100 to 120 steps per minute) will help reduce stress, heighten social connectedness, maintain baseline health, and improve posture. If you're relatively fit, it may not do much to *build* muscle, flexibility, or endurance, but the other benefits endure.[5] Walking up stairs instead of using elevators, parking in a spot farther away from your destination, walking the dog, and taking short, brisk walks to break up the day all add up.

3. *Calorie-burning mode.* At three to four miles per hour (135 steps per minute), you enter "exercise" mode. You increase aerobic capacity, and by burning eighty-five to one hundred calories per mile, you can shed pounds. "Telephone-pole walking," where you alternate between fast and aggressive walking every few hundred feet, is a good way to move up to the next mode.

4. *Muscle-building mode.* At more than four miles per hour (150 steps per minute), you're into aerobic AND muscle-building territory. You'll burn (slightly) more calories per mile and shred—thus build—muscle. Squeezing your butt muscles every ten paces, carrying weights, and vigorously swinging your arms can increase results.

Most people walk about four thousand steps in the course of a day. Some experts suggest that ten thousand steps per day (four to five miles) are necessary to maintain health; others suggest thirty minutes of brisk walking or other activity five days a week.[6] Unless you lead an extremely active work or home life, reaching these goals often involves setting time aside to walk. It's worth it.

Studies have shown that walking can produce a 20 percent reduction in the chance of having a stroke, a 30 percent lower risk of developing cardiovascular disease, and a 40 percent lower risk of fracturing a hip.[7] In women, there can be a nearly 20 percent reduction in the incidence of uterine and breast cancers.[8] And the payoff pitch: walking—per the Blue Zones mantra—can add up to seven years to your life.

Walkers can reap the same benefits and burn about the same number of calories per mile as runners—it just takes longer. But it's safer: the airborne gait of runners imposes one hundred tons of impact on legs and feet per mile; the risk of injury for walkers is 45 percent less than that of runners.

I talk with a representative from a local health-care provider helping sponsor the Salinas charrette. When I ask him "what's in it" for his company, he explains that 75 percent of the city's hospital population is under government care and the hospital gets reimbursed only 30 to 40 percent on the dollar. "We lose money when most people walk in the door," he explains. "Our idea is to keep you out of the hospital—and the only way to do that is to help people live healthier lives and make better choices. Walking is key."

Psychological (and philosophical) benefits. Walking is a natural antidepressant. The body's endorphins, released while walking, elevate one's mood throughout the day. Exposure to sunlight can reduce the incidence of seasonal affective disorder. Walking can bolster memory and improve sleep.[9]

It can also spark deep thinking and creativity. Immanuel Kant was so regular in his walks through Königsberg he was nicknamed "the Königsberg clock." He walked alone, believing one should breathe only through one's nose, and any companions might encourage talking through the mouth. Friedrich Nietzsche also walked alone, but he often perambulated for eight hours a day. He carried notebooks and plotted the courses of most of his great writings while on foot. He maintained, "Sit as little as possible; do not believe any idea that was not born in the open air and of free movement—in which the muscles do not also revel."

If you don't mind being gawked at, walking backward can improve cognition and help you think faster on your feet.[10] Walking can even make you more literate; Dr. Joseph Mercola—who walks barefoot on the beach about fifty-five miles every week—reads two to three books each week as he walks, though this is hardly ideal for people walking in traffic.[11]

The Monterey Bay Area Walking Club convenes twice a week to walk four miles and is open to all. Potlucks, wine tastings, and enduring friendships have evolved from the group.

Social benefits. One Tuesday afternoon, I arranged—through Meetup—to take a hike with the Monterey Bay Area Walking Club. The dozen people who gathered carried backpacks, walking sticks, water bottles, and leashes tethered to Zooey, Shiloh, Eiffel, and other canines. It was a cocktail party on foot, but the glue that bound this eclectic group together wasn't martinis; it was healthy companionship. The walkers wore tennis shoes, hiking boots, and flip-flops. Some strode quickly as if they were about to miss their bus, some sauntered, but most kept a brisk conversational pace. There were tall, short, skinny, not-so-skinny, old, and middle-aged legs. During our four-mile jaunt, we encountered rain, sunshine, fog, wind, thirty nice drivers, and one asshole.

I had a long chat with former *Californian* magazine editor Katharine Ball, who hiked her way to health twice, once recovering from a rare lung infection and a second time after battling ovarian cancer. "Walking saved me," she explained. "I don't mind walking alone, but as I get older, I fear falling and getting injured in some faraway place, so I find it more comforting to walk in a group."

Walking can be done by almost anybody, almost anytime, almost anywhere. All you need are shoes—and even those are optional. To reap maximum benefits, here are some suggestions:[12]

General posture. Tuck in your abs (to avoid arching your lower back), stand tall, and keep your shoulders back.

Legs. To increase speed, take quicker strides instead of long strides, which can hurt your lower back.

Feet. Start each stride with your front foot at a thirty- to forty-five-degree angle to the ground. Roll your foot forward, then to begin the next stride push off from the toes of your back foot. A person behind you should be able to see the sole of your shoe.

Arms. Hold your forearms at a ninety-degree angle to your upper arms and swing them naturally. Relax your shoulders. Keep your elbows close to your torso. Limit your arm range of motion so your hands don't go higher than the middle of your chest or farther back than your hip.

Eyes and ears. Focus on an area twelve to twenty feet ahead of you, or three to five sidewalk "squares." If you like walking to music or podcasts, consider wearing just one earbud or a single headphone so you can be aware of traffic and other sounds around you.

Shoes. While running shoes have a wide, thick heel to absorb initial impact and offer stability, walking shoes have a more uniform, flexible sole with better arch support that allows the foot to "roll" more easily.

Walking: It really is as simple as putting one foot in front of the other.

Stepping Out with the Nordic Walking Queen

"If walking is vanilla ice cream," muses Linda Lemke, "then Nordic walking is like adding hot fudge." And taking a walk with Lemke—the Nordic Walking Queen—is like adding a charming, red cherry on top.

Queen Lemke's castle is a quaint two-bedroom rambler on Schneider Lake in Cold Spring, Minnesota. Her scepters are two 110-centimeter-long walking poles. Her kingdom is anywhere she chooses. She became "queen" after one ill-disposed, yet eventually converted, participant in one of her walking classes whined, "Well, who made *you* queen?"

Nordic walking is nothing more complex than walking with a pair of specially designed poles. When you walk, you naturally swing your arms; by putting poles on the ends of those arms and pushing off, good things happen. While ordinary walking engages 60 percent of your muscles, Nordic walking engages 90 percent, because you exercise the upper body, arms, core, and back. Because more muscles are working, calorie burn is increased by 20 to 40 percent and jumps the heart rate by up to thirty beats per minute. It also helps build bone density—important for everybody as they age, but especially people with osteoporosis. The poles reduce the impact of each step and "turn you into a quadruped," as Lemke puts it, which decreases a person's risk of falling.

"The right poles are critical," Lemke explains. "If you went mountain biking for the first time on a road bike, you might think you hated it—but the real problem would be the wrong equipment. The same is true with Nordic walking." A pair of Nordic

The Nordic Walking Queen, Linda Lemke, walks the walk. Nordic walking burns 30 percent more calories, engages 30 percent more muscles, and helps build bone density—for about 10 cents a day.

walking poles costs between $100 and $200 and will last ten years or more. If you add the cost of replacing the $20 angled pole tips once a year, you wind up with an exercise regimen that costs 10 cents a day—a regimen you can do anytime and anywhere.

We start off on our three-mile walk. Initially, Lemke has me simply swing my arms naturally and drag the poles. The idea is to not think, just walk. As I find my natural rhythm, she has me grip the handles and plant the pole tips with more conviction. Half an hour into our walk she has me push off with my poles and my toes, "adding energy," as Lemke explains it. And I'm Nordic walking—more like a serf than royalty—but moving in the right direction. And it works—my upper body is surprisingly, yet pleasantly, sore the next morning. Lemke explains how people have tried to increase the benefits of walking by swinging their arms vigorously, wearing leg weights, or carrying backpacks, but Nordic walking is more natural and more aligned with how our bodies were built to move.

With more than ten million participants, Nordic walking is one of the most popular forms of exercise in Europe. Lemke is baffled, even a little peeved, that the sport is still so slow to catch on in North America. But she has a theory: "There's a certain geekiness factor when it comes to poles and Nordic walking, and Americans don't like to be geeky." She explains her neighbors' reactions when she and her husband began walking with poles. "People would roll down their windows and ask if we'd forgotten our skis or had been injured or were having problems with balance." But once people try it, they love it. After one introductory class involving mostly middle-aged women who worked together, nineteen of the thirty participants purchased poles. "Now they have walking groups that go out at lunch and after work. It's become their thing."

Nordic walking is used extensively in therapeutic and rehabilitation circles. Lemke has taught Nordic walking for nine years to people with Parkinson's disease. One Chicago hospital is using Nordic walking to help breast cancer survivors build upper body strength.

It's a superb exercise as people age, as it helps people maintain a healthy posture and balance and stay physically and mentally active. She cites Bob, a formerly active ninety-year-old whose health had declined to the point where he had moved to assisted living. His kids bought him a walker, which he roundly rejected. Instead, he began walking with Nordic poles. At first, he could make it only one hundred feet from one bench to the other as he walked around a nearby lake. But soon he was walking the entire two-mile shoreline. At ninety-eight, he's still Nordic walking.

"It helps if you catch people early," Lemke says.

She and I goof around with variations of Nordic walking. We skip, double pole, do tummy crunches, zigzag uphill, and walk in bursts. I note that drivers are polite—some give us a prolonged glance but most give us a wide berth. I ask Lemke about safety. "If I were a mugger, going after someone with poles wouldn't be my first choice," she says. "Dogs may bark, but they stay away because they're not sure what the poles are all about. And when we walk at night, we wear reflective clothing. I've never felt in danger."

As I leave, I discover the Nordic Walking Queen is also a Nordic doctor. She hands me a slip of paper with a prescription. The drug is listed as "walking," the dose is "thirty minutes per day," and the ancillary information says, "OK to repeat multiple times per day."

· 16 ·

THE BLOCK

The Feast *of a*
Great Community

———

ON A RECENT VISIT TO NEW YORK CITY, I ATTENDED A PERFOR-
mance of Blue Man Group—that mute trio of azure men who,
through their pantomime, drums, and marshmallows, bring our
shared fears and joys to the surface. One skit involved a deep-
voiced narrator discussing how we are all linked together, how we
are all ONE. A movie flashes on the screen, showing an escalating
collage of people, emotions, places, crowds, and events. With
bated breath, we await to see and hear the deep philosophical message that will reveal our "ONENESS." Suddenly, the camera dives
underground, weaving and bobbing to reveal the universal truth
that links us all together: sewers. Miles and miles of PVC pipe.

True, our blocks, neighborhoods, and cities are bound together
by sewer lines, concrete curbs, and wire, but there's something else.
What is it that makes one three-hundred-by-six-hundred-foot
block more pleasant to walk around than another? Why do some
neighborhoods beckon us not only to walk through them, but to
stay? Why are we, like the racing pigeons we'll soon visit, magically
drawn home regardless of how far we've roamed?

Designer Jan Gehl offers this thought: "A good city is like a good party; people don't want to leave early."[1] The simile is fitting. A successful city and a successful party involve the right setting, the chemistry of strangers, the comfort of friends, the smells, the music, the surprises, the spontaneity, the unspoken etiquette that prevents exhilaration from spilling into mayhem—things that are hard to define, difficult to put into words, and impossible to touch. How do we start giving shape to the elements that draw us to certain locales like two-year-olds to a mud puddle?

I can think of no better place to look than the pages of *A Pattern Language: Towns, Buildings, Construction*, an obscure, 1,171-page book most people would walk right past on the yard sale FREE table. Written in 1977 by Christopher Alexander and others from UC Berkeley, it espouses silly notions like helping your teen build their own cottage to encourage them to express the "beginnings of independence."

It's one of my favorite books ever.

It overflows with two of my most beloved writing elements: (1) nouns, and (2) the authors' ability to take a foggy notion and give it shape with words—to translate English into English. For once a foggy notion has shape, you can carry it somewhere.

Here are a few ideas from *A Pattern Language* that helped me realize why some blocks, neighborhoods, and cities feel so "right":

- **Great neighborhoods have an identity.** They're often geographically small and intimate, composed of maybe five hundred people and encompassing no more than a few blocks. They have their own personality or cultural vigor. Residents feel safe and know one another. One person expressed, "I feel my home extends to the whole block."[2]

Healthy neighborhoods have public gathering spaces, informal places for hanging out, that welcome diversity and unique activities—like this plaza in the St. Germain area of Paris along the Seine River.

• *Vibrant neighborhoods have common places, public squares, and greens space where people can congregate.* The authors think that 25 percent open space is a nice proportion. These spaces provide room for natural groups to naturally gather—a place where older folks sit and chat, another where dog lovers and dogs can hang out, another for yoga classes, maybe benches for quiet reading. Big wide stairs and sitting walls also magnetically beckon people to gather, chat, relax, and feel like they belong.[3] Great neighborhoods have few heavily trafficked through-streets, which makes the area feel contiguous; people aren't just "passing through."

- *Healthy communities contain the basic amenities people need within easy biking or walking distance.* Housing, retail stores, schools, churches, and services intermingle or are in near proximity. These communities can consist of a neighborhood, an entire city, or a group of districts stitched together; five thousand to ten thousand people is about the right size. At that scale individuals feel their voices are heard. "No citizen [should] be more than two friends away from the highest member of the local unit," Alexander maintains. Your friend's friend should be able to connect you with a council member or mayor so you can directly voice ideas and concerns.[4]

- *Right-scaled neighborhoods are populated by buildings that are four stories or fewer in height.* They're more human-scaled. People living "low" are more likely to spend time outdoors and intermingle with others. From three stories up people can easily distinguish faces. At ten stories, the scene below more resembles a video game; people feel detached. Residents in Glasgow tenements had a custom of flinging a slice of bread with jam to connect with those playing below; you can't do that from a hermetically sealed space one hundred feet up.[5]

- *Great neighborhoods have walking and biking paths, with intermediate points of interest.* People enjoy observing things as they commute versus just "getting there." Points of interest can be fountains, statues, promenades, food carts, gardens, playgrounds, architecture, street musicians, graffiti, gentle curves, or changes in elevation.[6] There are elements for all to enjoy. People are free to display elements that are an expression of who they are: murals on walls, public art, blocks where boulevards are planted with flowers instead of grass.

Neighborhoods and cities are of every shape, size, location, economic vitality, and height; one size does not fit all. But for me, understanding the authors' patterns and way of thinking helps me recognize the things in our neighborhood and town that feel "right," and provides me with clues for changing those things that don't.

Historically, there have been only two basic models of urban growth: the traditional neighborhood and suburban sprawl. While traditional neighborhoods in cities and first-ring suburbs are characterized by mixed use, walkability, and diversity, suburban sprawl—especially in the outlying exurbs—often involves segmentation: subdivisions of tract houses sit apart from shopping areas which sit apart from office parks which sit apart from manufacturing areas which sit apart from parks and city services. Car is king; ten to twenty trips per day go along with each suburban household. The authors of *Suburban Nation* liken American suburbs to "an unmade omelet: eggs, cheese, vegetables, a pinch of salt, but each consumed in turn, raw."[7]

Two post–World War II policies changed the way the United States grew and lived. First, Federal Housing Administration (FHA) and the US Department of Veterans Affairs (VA) housing programs provided mortgages for eleven million new homes—almost all single-family, suburban tract homes. This helped ease the postwar housing crunch. But since urban renovation projects weren't included, suburbs sprawled while inner cities crawled. The second policy was the building of the interstate highway system—which meant those people living in new suburban areas could reach the segmented parts of their cities easily—invariably by car.

Studies show that today, urban, suburban, and exurban areas are growing at roughly the same rate (rural areas are experiencing a downtick in population). Outlying suburbs aren't *bad*; they're just badly positioned to take advantage of the ideas in *A Pattern Language*. It's good to have choices; the question is whether or not you want to have the delightful option of your neighbor tossing you a slice of jam bread.

PART IV

NATURE

PIGEONS

Vindicating Rats
with Wings

———

NO PERSON IN HOLLYWOOD HAS DONE MORE TO MALIGN THE REP-
utation of an entire species than Woody Allen, who, in the 1980
movie *Stardust Memories*, refers to pigeons as "rats with wings." As
he chases a winged intruder through his apartment with a fire ex-
tinguisher, he sputters, "It's probably one of those killer pigeons."

Are pigeons as Allen proselytizes? Are they invaders or enter-
tainers? Urban panhandlers or symbols of peace? Bird brains or
geniuses? Well, it depends on who you ask.

Ask ornithologists and they'll tell you, of the 18,043 species of
birds on earth, 300 are pigeons or doves belonging to the Colum-
bidae family.[1] Ornithologists will explain that rock pigeons—the
blue-gray pigeons we not only encounter on our daily walk, but are
used for racing, as messengers, and as antagonists in Woody Allen
movies—were domesticated in the Middle East six thousand years
ago and brought to North America by French settlers in the early
1600s.[2]

Ask bird watchers and they'll tell you pigeons have habitats and
habits all over the map. Tumbler pigeons somersault through the
sky. Frillback pigeons appear to have lost a brawl with a curling

iron, and crowned pigeons are nearly the size of a turkey.[3] And there are an estimated four hundred million rock pigeons in the world, making them an easy checkmark in the box of any birder's life list.[4]

Ask building supervisors and they'll tell you that the twenty-five pounds of droppings a pigeon generates each year can block gutters, wreak havoc with rooftop ventilation systems, and create slippery stairs and sidewalks, especially when mixed with rain. They'll also tell you that the acid content can deteriorate stone exteriors and paint; that pigeon control for a high-rise can cost north of $100,000; that deterrents include spikes, shocks, optical gels, plastic owls, ultrasonic repellents, hawks, holographic tape, strychnine, cannon netting, birth control, predator eye balloons, and playing Myron Floren accordion music. They'll also tell you that relocating them doesn't work since pigeons—as we'll soon see— have an impeccable homing instinct.

Ask public health officials and they'll tell you several things. Most will concur that the droppings can provide a home for fungi that can lead to a lung disease called histoplasmosis; that nests can harbor parasites, ticks, and mites; that "bird fancier's lung," a type of pneumonitis caused by prolonged exposure to airborne dust from droppings, can be serious—but that all the above are true of many birds. Most will grudgingly admit there is little evidence that pigeons carry diseases that directly affect humans.[5]

Ask philoperisterons—pigeon lovers—like Tina Trachtenberg of New York City, and she'll tell you, "To me, pigeons are little beds of flowers as I walk around the city."[6] Ask pigeon fanciers Mike Tyson or Queen Elizabeth, or past pigeon fanciers like Elvis Presley, Walt Disney, Charles Darwin, Roy Rogers, or the "Queen of Strippers" Gypsy Rose Lee, and they'd sing their praises.[7] If you could ask Nikola Tesla, father of alternating current—who would bring sick pigeons to his hotel room to nurse them back to health—

Dr. Phil Nelson with some of his 240 pigeons, one of which won a recent six-hundred-mile "old bird" race. "They're superior athletes," Nelson says.

he'd tell you, "I loved that pigeon as a man loves a woman, and she loved me. As long as I had her there was a purpose to my life."[8]

Ask Reuters reporter Alan Cowell and he'll tell you they're indispensable. Without radio or phone in the 1980s, he used pigeons to deliver news about Zimbabwe's struggle for independence from a remote area of the country.[9] Go back 130 years and ask Reuters founder, Paul Reuter, about how he used pigeons to transmit stock market quotations between Germany and Belgium, covering the distance four hours faster than a train.

Who to ask?

I decide to ask Phil Nelson, a semiretired veterinarian who has spent most of his life keeping pigeons and the past forty-five years racing them. I invite myself to shadow him—and his 240 pigeons—over the course of a week. An unbiased source? No. Interesting? Very.

I'm chatting with Nelson on the upper level of his coop as he snatches forty-five pigeons from their enclosure and places them in two carrying crates; he has the swift, sure hands of a seasoned shortstop. I clutch one pigeon as Nelson fiddles with a leg band and can only compare it to holding an oven mitt with restless leg syndrome.

We're about to take his "young pigeons"—four or five months old—on training runs. We place the crates in the trunk of his car and head out. Nelson explains that his pigeons are descendants of the common rock pigeon we see every day. But he also explains that if you set a feral rock pigeon next to one of his, you'll notice his are more muscular and have wider chests—the result of selective breeding. "They're superior athletes," Nelson crows.

The pigeons' training regimen began weeks ago. He let his pigeons loose in the expansive yard around the coop, allowing them to explore the area and then find their way back to the coop. Then, he released ("tossed") them from a block away. Next will be a mile, then ten, then forty. After that, they'll be ready for their first hundred-mile race. Nelson and I drive a mile, open the first crate along a rural road, and watch the birds explode upward like a barrage of sky rockets. We drive another thirty miles, this time to a parking lot, where Nelson releases a second group. "They'll all beat us home," he states with complete confidence. They do.

How?

"They're racehorses of the sky," Nelson explains. "They're the toughest birds on earth and extremely intelligent." Though a pigeon's ability to find its way home isn't completely understood, even by researchers at Cornell University who have studied them for years, many of the pigeon's unique abilities are known.

"They can see fifty miles," Nelson says. "Farther than an eagle. They use the position of the sun to navigate. And they have a superior sense of smell." They can also see the ultraviolet part of the light spectrum invisible to human eyes. Nelson tells me about two other attributes that contribute even more to the pigeon's uncanny homing abilities. "Their hearing is so sensitive they can detect the wind howling over the Rocky Mountains from a thousand miles away." Then there's their supernatural sensitivity to the gravitational and magnetic forces of the earth. It's a sixth sense—in the truest sense— that pigeons and other animals have. As someone who can get lost eight blocks from home, I surely lack this sixth sense.

They're smart—or at least easily conditioned. Pigeons have been trained to distinguish between paintings by Monet and Picasso and music by Bach and Stravinsky.[10] They can recognize all twenty-six letters of the alphabet.[11] The psychologist B. F. Skinner, who headed the military's Project Pigeon during World War II, showed that a pigeon housed inside a warhead could guide a torpedo by pecking at the image of a ship in a specific window. He also taught the birds to play ping-pong. Being called a bird brain may not be that much of a slight after all.

"Most likely," Nelson explains, "their homing ability is a combination of hearing, sight, smell, memory, and sensitivity to magnetic forces." Nelson should know; the "old bird" races held earlier in the summer culminated in a six-hundred-mile race. One of his pigeons won it.

Given that pigeons were originally remote-cliff dwellers, it seems odd they've become such a staple of urban life. This is partly due to their proclivity to perch and nest in the relative safety of high vertical places—be it cliff, railroad trestle, or city hall window. Another factor contributing to their wide range of habitats is their social nature—they love to congregate with one another and, when they deem it safe, with humans. And, like squirrels, they have great culinary adaptability; they're happy eating a corn dog in the suburbs, popcorn in Central Park, or feed corn in rural Iowa.

They breed for life ("or until their mate has been away for two days," Nelson quips) and share equally in chick-rearing duties. Both parents feed chicks by regurgitating a milky substance from their neck crop into youngsters' mouths.

Despite their metropolitan sensibility, pigeons are not particularly gifted architects or housekeepers. They build flimsy nests and often reuse them year after year. They don't carry away waste like many birds, which is why their nests resemble something out of *Hoarders*, containing candy wrappers, old eggshells, and even mummified nestlings.

Humans have a long-standing relationship with pigeons; in days gone by, if you provided a dovecote for nesting, pigeons would provide you with eggs for breakfast and squab for dinner. They also provided guano, a rich form of fertilizer for fields. That same guano provided the saltpeter needed for explosives. But perhaps their most important contribution has been providing the very earliest form of wireless communication.

Domesticated pigeons were used thousands of years ago by captains of incoming Egyptian ships to announce their impending arrival.[12] Early Romans, Genghis Khan, and Charlemagne—they all established pigeon networks for communicating. Indeed, up until Samuel Morse invented the telegraph in 1844, messenger pigeons were by far the quickest form of communication. Unaided pigeons can fly, on average, forty miles an hour. Aided by a tailwind, they can reach eighty miles per hour or faster.

Yet using pigeons for communication is a thing of neither fantasy nor the distant past. In October 1918, 550 men of the 77th Infantry Division were trapped behind enemy lines near Verdun, France. Radio communication was down, and the runners the division's major had dispatched had all been killed or captured. So pigeons were sent to communicate the dire situation. The first and second birds were shot out of the air by German forces. In a third attempt, a note was slipped into a canister attached to the leg of pigeon Cher Ami that read, "We are along the road parallel to 276.4 . . . our own artillery is dropping a barrage directly on us. For heaven's sake, stop it!"

Cher Ami was shot almost immediately but recovered and winged her way twenty-five miles to division headquarters. When she arrived, they discovered she'd been shot through the breast, blinded in one eye, and had one leg hanging by a single tendon. But the friendly fire was halted, and Cher Ami helped save the lives of the remaining 194 soldiers. Based on her heroics during this and other battles, Cher Ami was awarded the Croix de Guerre medal for valorous service. Thirty-two other pigeons have been awarded the Dickin Medal, a sort of Victoria Cross for animals, for wartime bravery—an excellent showing compared with the eighteen dogs, three horses, and one cat that have also been awarded the medal.

Nelson explains that during the Vietnam War the US Coast Guard trained pigeons to spot the bright-colored lifejackets of downed pilots. Pigeons were placed in glass globes beneath rescue helicopters; when one saw the jacket of a pilot, it would peck at a signal target to inform the flight crew. The birds' eyesight was ten time sharper than that of the helicopter crew and could often locate a downed pilot faster than emergency radio transmitters. Nelson further explains how during the Gulf War, the army shipped thousands of trained pigeons to Iraq to carry messages between outposts in the event of a nuclear attack, when wireless communication would be impossible.

On the dark side, drug traffickers have used pigeons to transport heroin between Afghanistan and Pakistan, ten grams at a time.[13]

Not all has been lovey-dovey in humans' interactions with pi- geons. When Europeans first landed in North America, passenger pigeons—indigenous to the area—were the most abundant bird on the continent, if not the planet. In 1860, naturalist W. Ross King encountered a flock he estimated to be one mile wide and three hundred miles long; some people guess the flock contained two to three billion birds.[14] Many took advantage of this seemingly inexhaustible natural resource. When drought and pestilence wiped out crops and livestock in the 1700s and 1800s, pigeons were a free, often lifesaving source of food for human and beast.

Passenger pigeons traveled so quickly—up to sixty miles per hour—and in such vast numbers that many mistook the approaching beating of wings for tornadoes or the apocalypse—and for some rural dwellers it *was* Armageddon. When descending en masse to roost for the night or to forage, pigeons could decimate entire forests by toppling trees and breaking limbs.[15] The largest flocks could consume fifteen million bushels of food in a single day. If you were a farmer who had just planted or were ready to harvest corn or wheat, you faced devastation.

Because many people began seeing the passenger pigeon as the enemy, slaughtering them became a mania. Flocks flew low and were slow to react; they could be swatted out of the air with oars, clubs, stones, or even potatoes. A single shotgun blast could kill a hundred birds. They were netted in batches of thousands. "Stool pigeons"—live pigeons tethered to a sort of teeter-totter that made them look like they were landing—were used to lure pigeons toward netting areas. Once netted, the birds were bludgeoned to death or captured live and used for one of the more popular sports of the day—pigeon shoots.

Destruction of habitat was their final undoing. Wetlands, a vital source of food, were drained, and forests were set afire. Nesting sites were obliterated, including one where fifty thousand birds were killed every day for five months.[16] Deforestation—hundreds of millions of acres of forest were cleared for agriculture and lumber production—

eliminated roosting habitats and the bird's favorite food, acorns.

Soon, but not soon enough, people realized that passenger pigeons had been slaughtered to near extinction. Legislation and conservation measures failed to halt the unthinkable. Attempts were made to net birds, this time to *preserve* the species. By 1900, only three small breeding flocks were left—all in captivity. A few eggs were hatched, but by 1910, only a single passenger pigeon remained—a female, named Martha, kept at the Cincinnati zoo. She lived a life of solitude for four years and then died on September 1, 1914.

When people saw flocks of a billion passenger pigeons passing overhead in 1860, they were dumbfounded by the inexhaustible numbers. Fifty years later, when the last one died, people were bewildered by how quickly one species could eradicate another.

Pigeons learn fast; when this woman rode into a square in Rome—though there were dozens of other bicyclists around—hundreds of pigeons flocked to meet her (and her bag of food) for their morning ritual.

The Friday before race day, I spend an evening registering pigeons with Nelson and other members of the Minneapolis Racing Pigeon Club (MRPC) at their clubhouse inside an airplane hangar at Anoka County Airport. We stomp our feet on a rubber WELCOME mat soaked in disinfectant and then carry in two crates of pigeons for the "100 miler." Drifting in and out of the hangar are a dozen people—all guys—ranging in age from thirty-four to eighty. Included in the group are computer programmers, heavy-equipment operators, veterinarians, and retired welders. The area is stacked with shipping crates, and its walls are decorated with pigeon posters. The floor is crowded with folding tables holding electronic registration equipment.

Registration requires four people: one person uses a computer to tabulate the process; a second person—the bird owner—removes the bird from the crate and hands it off; a third reads aloud the band number on one leg before swooping the other leg—this one fashioned with a microchip band—over the scanner to encode the secret number for the race; and a fourth places the bird in one of a dozen release pens.

On race day, the pigeons will be simultaneously released from a special trailer and wing their way to their home coops. Upon entering the home coop, the number encoded in the leg band microchip will be recorded by a module. Since different coops are different distances from the release point, the winner will be determined by calculating the speed of each bird. It's the sport with "a single starting gate and a thousand finish lines."

Timing is everything. Nelson lost one three-hundred-mile race by three-eighths of a second. The current computerized system is a far cry from previous ones when pigeon racers needed to grab their birds and run to the Town Square, where someone verified the arrival time with the town clock, or the not-so-long-ago method of placing a rubber band from an arrived pigeon in a slot of a windup clock.

While the seven hundred pigeons are being registered, the men find plenty of time for banter. Storytelling intermingles with the cooing of pigeons, the yelping of registration numbers, and the roar of airplanes.

"Number nineteen, zero, one, two!"

"Ya know, one company used to use pigeons to send blood samples to the lab for testing. It was the fastest way to do it and they saved $60,000 in courier fees."

"Number one, oh, seven!"

"That pigeon on the wall, that's G.I. Joe. Damn bird flew twenty miles in twenty minutes to carry a message that called off an air strike that woulda killed a thousand US soldiers."

Coo, coo, coo.

"Number nine, thirty, two!"

"Hey, how do we know Spike ain't some undercover PETA agent?"

"He better not be. We'd have to tar and feather him."

"Number seven, seventeen!"

"Yah, we got plenty of feathers."

"Number eight, oh, nine!"

"This sport is dying fast. If it weren't for the Asians—we got three of 'em—this sport would be dead."

Putta-putta-putta-putta-vvvvvrrrrrroooom.

"Yeah, in Europe and Asia, the sport is booming. Some Chinese guy just spent over a million bucks for a Belgian racing pigeon."

"Number forty-two, eighteen!"

"But you know, everybody gets along. It doesn't matter what color you are. Who gives a shit? You just fly your pigeons."

"One, nine, nine!"

"There's so many different kindsa people that race. If it wasn't for the pigeon club, I'd probably never talk to any of these people."

"Okay, what's the longest it's taken for someone's bird to come home?"

Coo, coo, coo.

"My son had one that was—what was that, John, six years? When it finally entered the coop, we sprayed it with a garden hose so we could catch it and read the band. Six damn years!"

"Who wantsta bet this new cloud system for clocking the birds isn't gonna work worth a shit? I'll give anyone five-to-one odds it won't."

One new club member talks about losing fifteen pigeons to an unknown predator on a single night. The man figured out a mink was the culprit, because mink like to stack their prey like cordwood. Another guy explains chasing birds is easier than chasing girls because the former are less expensive.

When the seven hundred pigeons have been registered and crated, we load the dozen pens onto a flatbed trailer and then haul them ten miles to a release trailer, where they're loaded along with pigeons from three other clubs. When all the crates are ready to go, Tom—the truck driver and "liberator"—closes the doors and drives one hundred miles through vacation traffic to Alexandria, Minnesota, where he'll release the pigeons the next morning.

Most of the pigeons, I'm told, will find their way home. Those that don't will wind up in other coops, living with "scrubs" (feral pigeons), or, more rarely, run into wires or be picked off by a peregrine falcon. But what motivates them to fly their tail feathers off?

Nelson says many are simply born to race. "They love what they do, especially the older ones. When they're crated up, the old pros will go sit in the corner to conserve their energy while the young ones are up all night pecking at one another, acting like teenagers. In the morning, the old guys are off and running. They know what they're being asked to do. And when they come home, they don't hit the roof right away. They'll make a couple of victory laps and pop their wings like they're giving you a high five. They're just a joy to work with."

Most pigeon racers use some type of motivator to increase the urge their pigeons feel to beat a beeline home. Some use the "widowhood" method. Pigeons mate for life, so some breeders will separate a racer from his or her mate for a few days, and then let them see, but not become amorous with, one another before sending the racer off to the race venue—arousal through pigeon porn.

Others fly their pigeons "to the nest," which involves timing birth cycles to remove a racing bird from the nest when the chicks are seven to ten days old. While one pigeon is away racing, the mate takes over the parenting duties; but still, the removed bird will beat a hasty retreat home to satiate his or her parenting instincts.

Other racers use jealousy as a carrot. Pigeons are conservative; they don't like change and prefer sticking to a single nesting box. Some owners will place a new pair of pigeons, or even a mirror, in a nesting box to bring out the feistiness of the displaced pigeon before a race.

Others use food as a motivator. Most racing pigeons are unpracticed in foraging for food and will hasten home for a good home-cooked meal—even when that home-cooked meal isn't cooked, it's six hundred miles away, and they've flown over hundreds of miles of perfectly edible corn along their way.

Some simply use the familiarity, safety, and security of the coop as a motivator.

One could also question what motivates the humans who race the racing pigeons. In most cases, it's not the prize money. Though some races, like the South African Million Dollar Pigeon Race, promise a first-place purse of $200,000, most payouts amount to little more than chicken feed. Gambling is a large component of racing in some cultures and countries, but at least in the MRPC, this seems to be a non-factor.

More than once I heard the word "addicting" ascribed to the sport. There is clearly satisfaction in breeding and raising birds from birth, then tossing fate and birds to the wind to see whether you've done your job better than anyone else. Many consider pigeons the poor man's racehorse.

"Raising pigeons is the hardest thing I've ever done," Nelson told me, "and I made it through vet school and raised one and a half-dozen kids." (Think about it—he means seven.) He's so passionate about pigeons that thirty-one years ago, he found himself in his backyard in a tuxedo anxiously waiting to clock in a pigeon so he could get to his wedding on time. When I ask his wife what it's like to be married to a guy with 240 pigeons, she replies with a loving grin, "Lonely."

I rise at 4:30 a.m. on Saturday and drive the 120 miles to Alexandria for the pigeon release. The trailer is set up, and Tom the liberator pulls the lever. The birds bolt for the sky. It's an alternate-world tornado, one that swirls upward rather than down, one that makes you stare in utter awe rather than terror, one that's gentle and fluttery rather than fierce, one that makes you wish it would sweep you up and carry you away.

The pigeons gradually circle upward, testing the wind currents for an easterly direction. Latecomers join the flock. For a few seconds, they swoop back toward the trailer, and then, as if by signal, they head home. They're driven by lust, by love of doing what they were born to do, by the urge to parent, by hunger, by a longing for home—like us.

HACKS & FACTS

Why Pigeons Strut

When pigeons seemingly strut and bob their heads, they're actually holding their heads stationary for an extra split second while their body moves. This visually stabilizes the world around them for an extra second to increase the time their eyes and brain have to identify predators or dangers. They're not trying to be cool; it's an act of self-preservation.

PARKS

Where Cities Pause *to* Catch Their Breaths

IF YOU STEP OUT OF YOUR FRONT DOOR IN SCOTTSDALE, ARIZONA, you have an impressive 40 percent chance of finding a park within a ten-minute walk of your WELCOME mat. But Kansas Citians have a 69 percent chance, Clevelanders an 82 percent chance, and San Franciscans a 100 percent chance—an absolute certainty—of being able to wiggle toes in park grass within six hundred seconds of walking out the door.[1]

Overall, 70 percent of Americans are fortunate enough—and urban enough—to live within a ten-minute walk of a park. But is it a *great* park? And if so, what makes it *great*? In its annual Park-Score survey, the Trust for Public Land grades a hundred major city park systems on walkability, per capita expenditure, acreage, and amenities, including the number of basketball hoops, dog parks, community gardens, playgrounds, and Pickleball courts.

In 2019, Washington, D.C., St. Paul, and Minneapolis came out No. 1, No. 2, and No. 3, respectively. At the bottom of the list were Oklahoma City; Charlotte, North Carolina; and Mesa, Arizona. Minneapolis city park visitors, according to the Trust's Park-Score, can enjoy 124 restrooms, three disc golf courses, and 147 basketball hoops. Minneapolis spends $350 per resident on parks—a far cry from the $46.70 Oklahoma City spends—and dedicates 15 percent of the city's acreage to parkland, nearly five times as generous as Mesa's 3.5 percent.

But numbers answer only part of the "great park" question. There's something personal and idiosyncratic about the ingredients that go into a great park—something a ParkScore just can't measure. Let's hop on our bikes and pedal twenty miles to Minneapolis while we create our own scorecard to grade some of those unquantifiable elements.

"Grand Rounds" is fifty miles of winding asphalt that serves as the connective tissue for many of the lakes, rivers, playgrounds, ball fields, and historic districts that make up the Minneapolis park system.

We start at the Chain of Lakes—five bodies of water nicknamed by a nineteenth-century admirer who compared the interconnected waterways to "a necklace of diamonds in a setting of emeralds." Today, this jewel includes jogging and biking paths, beaches, band shells, dog parks, archery ranges, bird sanctuaries, rose gardens, sculpture parks, and seven million visitors. Over the years, the waters have welcomed canoes, sailing regattas, pond hockey tournaments, standup paddleboards, milk carton boat races, triathlons, a floating library, and fifty-one-inch muskies. VARIETY—check.

Just as community developer Andrés Duany once posited that "great cities are nothing more than a series of villages artfully stitched together,"[2] great park systems are little more than people and places sewn together by grass, activities, and mutual respect. We find zones where parents with toddlers congregate, areas for people who want to be left alone to read and sunbathe, spaces for those who want to mingle, areas for teens who want to play Hacky Sack, isolated benches for those falling in or out of love, areas where chess players assemble. It's all kept in balance by an anarchy of the most respectful type. It's the social contract in flips-flops. SELF-RULE—check.

Nature is abundant. As if early park foresters were prescient, many of the four hundred thousand trees in the park system are spaced the perfect distance for slack lining, hammock hanging, and shaded picnic areas. We walk our bikes through the Rose Garden, which boasts 250 varieties of roses, including those named after Barbra Streisand and "The Divine Miss M." We meander through the dwarf conifers of the adjacent Peace Garden and witness weddings, birthday parties, and photo shoots. We find signs telling us that many of the gardens, trees, and paths are tended by volunteer hiking and gardening clubs. We pedal past the "elf tree," an ash tree with an eight-inch-tall door in front of a hollow trunk that's home to the legendary "Mr. Little Guy." He dutifully responds to every message children leave behind—"My parents are getting a divorce," "I made this minnow cookie for you"—and ends every response with "I believe in you." COMMUNITY ENGAGEMENT—check.

Like most great park systems, the Chain of Lakes wasn't cobbled together as an afterthought but created as warp and woof of the city. But there's perhaps no better example of park forethought than New York City's Central Park.

Considered to be the first purposefully landscaped park in the United States, Central Park is an icon among icons. It has nine thousand benches, twenty-five hundred trees, thirty-six ornamental bridges, fifty-eight miles of walking paths, the wide-open fifteen-acre field of Sheep Meadow, twenty-one playgrounds, a thirty-five-hundred-year-old Egyptian obelisk, two million pieces of art (in the Metropolitan Museum), and—though now deceased—Gus the polar bear, who was treated with Prozac to deal with boredom neurosis, despite having had twenty million visitors during his life-time.[3] And despite being smack dab in the middle of the largest city in the United States, Central Park offers room for every body, every thing, and every activity. ELBOW ROOM—check.

Central Park has a large history too. Between 1821 and 1855, the population of New York City quadrupled, prompting the call for the early creation of a large central park. In 1853, acquisition of land between 59th and 106th Streets, bounded by Fifth and Eighth Streets, ensued. Imminent domain was declared, resulting in the eviction of sixteen hundred people, including residents of Pigtown and Seneca Village, a community of free African Americans. Land-acquisition costs for the eight-hundred-some acres totaled $7.5 million—exceeding the purchase price of Alaska's 424 million acres a few years later.

A design competition for the park was held, with Frederick Law Olmsted and Calvert Vaux—an English-born architect—beating out thirty-one other entrants. Certain stipulations needed to be met: the park was to have a world-class fountain, a parade ground, an exhibition hall, a lookout tower, and a skating area. Provisions needed to be made for four cross streets—ingeniously designed by Olmsted and Vaux to be recessed out of sight, eight feet below park level. Three million cubic yards of earth and rubble were shuffled around; thousands of truckloads of topsoil from New Jersey were hauled in. Eight ponds and lakes were formed by digging and blasting hollows that filled via natural springs.

Today, the park hosts forty-two million visitors a year, paupers and kings alike. In a recent census, four people listed Central Park as their home address. A beleaguered President Barack Obama once wistfully expressed, "I just want to go through Central Park and watch folks passing by. Spend the whole day watching people. I miss that."[4]

Back in Minneapolis, we continue biking around "the Lakes" and notice wetlands along the edges of three, areas that naturally filter storm runoff from nearby streets, lawns, and neighborhoods. Pollution and water control are two of the unsung attributes of parks. It's estimated that Philadelphia is saving $8 billion by using parks to manage storm water runoff and flooding rather than building an infrastructure of pipes and tunnels.[5] The City Parks Alliance maintains that urban park trees remove more than seven million tons of toxins from the air each year at a value of $4 billion.[6] The vast green spaces of parks help cool our cities. POSITIVE ENVIRONMENTAL IMPACT—check.

There are joggers, bikers, walkers, swimmers, in-line skaters, and canoeists at every bend. Average Americans spend 93 percent of their time indoors. But when people have ready access to parks, they are more likely to exercise. Something as simple as a walking loop can increase park use by 80 percent and attract twice as many seniors. Some people suggest we allocate some of the $10,000 annually per person spent on health care to supplement the $83 cities now allocate on average per person for parks. And just getting outside to exercise can be a powerful antidepressant—a "drug" many doctors are now prescribing. IMPROVED PHYSICAL AND MENTAL HEALTH—check.

Dogs—the ice breakers, the social magnets that draw strangers together—are the starting point for impromptu conversations everywhere; evidence shows that dog owners are five times more likely to get to know people in their communities than others.[7] And it's clear from the mix of those playing in pickup soccer and volleyball games that parks offer a common space that draws people of every economic, social, racial, and religious order together. LEVEL PLAYING FIELD—check.

From the Lakes, we chug over to the Minneapolis Sculpture Garden, where we marvel at the thirteen-foot blue rooster and plump Henry Moore sculptures. We find kinetic sculptures plunked in fields of natural prairie grasses, and plants that attract bees, birds, and butterflies. We jump for joy at the whimsical Spoonbridge and Cherry fountain. PUBLIC ART IN THE GREAT OUTDOORS—check.

We head for a well-earned cup of coffee at the Loft Literary Center. In front, we find a strange little oasis of calm on the sidelines of bustling Washington Avenue. It's a parklet

The author jumps for joy as he bikes through the Minneapolis Sculpture Garden, one of 266 unique parks in the city's park system.

carved out of three parking spaces cordoned off by shrubs and low walls containing benches, café tables, and inspirational quotations.

The parklet idea, born in San Francisco, has been so successful, the idea has spread to hundreds of cities worldwide. They're usually built and paid for by businesses, community organizations, or residents but are overseen by city planning or parks departments. Many are in close proximity to coffee houses, restaurants, or small shops and incorporate greenery, art, café tables, play areas, and workspaces. They're open to all and encourage walking and biking and provide a place for neighbors to interact. One Minneapolis parklet is used by a council member as an open office to engage constituents.[8]

This Minneapolis parklet takes up three parking spaces but in return provides a small public oasis of green and calm for gathering, resting, and eating. It will be dismantled in the fall to make way for snowplows and parking and then resurrected in the spring.

Triceratops topiary, community gardens with free spices for the snipping, children's events, and musical performances abound. Some parklets are made out of shipping containers or old Citroën vans brought over from France.[9] Some cost as little as $5,000, others more than $100,000. Like all good parks, parklets contribute to that blend of vibrancy and quietude cities need. *Architectural Record* pegged parklets as "the ultimate revenge on the modern city; one less parking space, one more park."[10] WACKINESS AND CREATIVITY—check.

We continue biking to the half-mile-long Stone Arch Bridge. Built in 1883, the bridge had fallen into disrepair, much like the surrounding area. But in the early 1990s, the bridge was transformed into a pedestrian and biking bridge—a key component in revitalizing the entire waterfront area. Parks have a real impact in transforming neighborhoods. The High Line in New York City, created from a mile-long section of abandoned elevated railway, hosts five million people a year. In Denver, $1.2 million in federal park grants spurred an estimated $2.5 billion in private and public investments. NEIGHBORHOOD REVITALIZATION—check.

Oakland Cemetery in Atlanta is equal parts park, cemetery, history center, sculpture garden, and event venue. "If you can remove the stigma behind cemeteries, they provide so many opportunities for community involvement," says Marcy Breffle of the Historic Oakland Foundation.

To discover what people did before the establishment of parks, I visit an unlikely place: Atlanta's Oakland Cemetery. Before the nineteenth century, there were few parks as we know them today. So people looking for a place to picnic, stroll, or engage in an occasional carriage race often headed to the well-manicured grounds of nearby cemeteries.

As I enter Oakland Cemetery, I pass a gaggle of touring Segways and a group of students sitting under a tree listening to their history teacher. Brick paths wind through tidy gardens, mammoth trees offer shade, and park benches beckon. A sign promotes the upcoming "Run Like Hell" 5K race. And there are dead people everywhere—seventy thousand of them.

"It's sort of like a miniature city," says Marcy Breffle, education manager of the Historic Oakland Foundation. "We refer to those buried here as residents."

Established in 1850, Oakland was part of the early nineteenth-century Victorian Garden Cemetery movement. It followed the lead of garden cemeteries like Père Lachaise Cemetery in Paris, where sculptures vie with headstones for attention and the tradition of leaving lipstick kisses on Oscar Wilde's tomb endures. In the United States, "rural cemeteries," more parklike in their look and feel, began transforming people's images of death from one of grimness to one of tranquility and acceptance.

History imbues the fabric of Oakland Cemetery. Golfing great Bobby Jones's tomb can be found with golf balls lovingly piled around it. The grave of *Gone with the Wind* author Margaret Mitchell is the most visited. The cemetery is the final resting place for sixty-nine-hundred Confederate soldiers and sixteen Union soldiers and includes three Jewish burial grounds, a Potter's Field, the African American Grounds, and dozens of regal sculptures and mausoleums.[11]

During my time with Breffle, she refers to Oakland as a cemetery and a city—and an outdoor museum, a history center, an arboretum, a sculpture garden, an event center, a community gathering place, and, of course, a park. The Historic Oakland Foundation hosts Sunday in the Park picnics, an annual "Tunes from the Tombs" musical event, and dozens of weddings and private events. When I ask Breffle whether people are ever taken aback by the intermingling of death and joy, she winks and jokes, "Well, we never get any pushback from the residents." Then she adds, "If you can remove the stigma behind cemeteries, they provide so many opportunities for learning and community involvement. No one ever wants their family to be forgotten—and what we're doing here is sharing the history and stories of our residents." EDUCATIONAL OPPORTUNITIES—check.

And what does the future hold for parks? Galen Cranz, a former professor at UC Berkeley, feels they may lead the way in showing people how to create a healthier environment. Community gardens, sustainable forests, mulching, and recycling centers could all be front and center.[12] I fantasize about parks becoming libraries for the body and soul—places where "park-arians" help you check out running shoes, bikes, and tennis rackets. Where people could renew not books, but arms, legs, lungs, and minds.

We circle back to my hometown of Stillwater and sear the brakes as we zoom downhill to Teddy Bear Park. Once a scrap metal yard, the park is unique with its twelve-foot-tall granite bears, river-themed playgrounds, amphitheater, and best of all— one feature few parks have today—the element of risk. There are climbing walls, rope spider webs, walkways with rabbit holes, and cargo nets for jumping into. And kids love it.

This kind of risk is an element built into many London parks. London parks cost a third less than those in the United States yet have 55 percent more visitors, of a broader age range. London's success is based on three factors: parks are designed for all ages, playful elements are scattered about (rather than centralized), and structures are riskier—without being dangerous. Researcher Meghan Talarowski explains, "When grandma is climbing three, four sets of stairs over and over again to go on slides, you know there's something special happening." She goes on to explain that kids yearn for the "Woo!" sensation of risk—and if it's not available outside, they'll stay inside and get their thrills vicariously on their smartphones or Xboxes.[13] RISKY POSSIBILITIES—check.

As we end our ride, it's fair to ask one another whether we can justify the economic, geographic, and manpower strains of urban parks—parks that often occupy the choicest real estate in a city. The land under Central Park alone is valued at a half trillion dollars. The annual parks and recreation budget of the one hundred largest US cities is a staggering $7 billion.[14] Still, as Mitch Silver, commissioner of the New York City Department of Parks and Recreation, explains, "I can't imagine a great city that doesn't have a great park."

And "great" comes in all shapes and sizes.

LAWNS

Caring *for* Your Seven Million Little Plants

WHEN I REACHED THE AGE OF TWELVE, THE PERFECT GREEN storm hit. My parents bought the suburban house of their dreams, surrounded by the lawn of their dreams. Included with the house was a lawnmower of nobody's dream. Having reached puberty, I was tasked with using said mower to mow said lawn.

Our lawn was the antithesis of those in Scotts fertilizer commercials, an obstacle course of hills, oak trees, roots, downspouts, and wavy-edged gardens. I could mow an occasional straight line, but my efforts mostly involved wrestling the mower into some form of hairpin turn while waging an uphill battle. The mower was a monstrous thing—part stump grinder, part Brahman bull. It made me loathe yard work. Forever. Today, our yard looks like something maintained by Thing One and Thing Two under the watchful eye of the Cat in the Hat.

Not so for Keith Smith of England, 2018 winner of the international Allett mower Creative Stripes Competition. His winning pattern mimicked that of an Amish quilt, and his previous year's pattern, a Union Flag. The pattern required mowing with military precision—two or three times a day. In the months before the

Keith Smith of Birmingham, England—2018 winner of the Allett Creative Stripes Competition—spent twenty hours a week wielding two 1940s push mowers to create this masterpiece.

COMPLIMENTS OF ALLETT

competition, Smith—a golf course groundskeeper by day—spent 273 hours on his lawn, more time than I spend on mine in five years.[1]

What's behind such green passion?

I load up my questions and truck and drive to the O. J. Noer Turfgrass Research Facility at the University of Wisconsin in Madison. I meet Paul Koch and Doug Soldat in the middle of a twenty-five-acre patchwork of experimental grasses. Three robotic mowers zigzag around us like caffeinated water bugs. As we talk, it becomes clear that Koch, a plant pathologist, and Soldat, a soil scientist, are the Rodgers and Hammerstein of turf.

I learn turfgrass isn't a weed, reed, or flower; like bamboo, corn, rice, wheat, sugarcane, and ten thousand other hollow-stemmed plants in the Poaceae family, it is taxonomically classified as, well, a grass. If we didn't have the grass family, we wouldn't have bread, beer, or Quaker Oats. Grasses are the most important source of food in the world.

Soldat pulls out a pogo stick–like tool, jumps on it, and from the tip pulls out a neat cross section of turf the size of a pack of cigarettes. He pulls the clump apart until he holds a single grass plant. "The essence of a turfgrass plant is the crown," he explains. He holds the plant so we can see the rice-size crown, with a small tangle of roots extending from the bottom and a sprig of green emerging from the top. "The crown sits at or just below ground level, and five to nine grass leaves [we call them "blades"] grow up from it. It's like a fountain with the newest leaves coming up from the center and the older leaves dying and falling away along the edges." A healthy lawn contains about eight grass plants per square inch—meaning your fifty-by-one-hundred-foot lot has 7,640,000 plants for you to tend.

Koch explains that the life cycle of any single grass leaf is only two or three weeks, and the crown rarely lives more than eighteen months. "Just as your body replaces cells so you don't have the same bones you had ten years ago," he says, "a lawn replaces cells so you don't have the same grass you had a year or two ago." Since the grass is mown before it has a chance to go to seed, turfgrasses reproduce via underground rhizomes that spread to form new plants.

Koch and Soldat and I wander around the research fields where seventy-five different studies are under way. We wander past plots of Kentucky blue grasses (which are neither blue nor native to Kentucky, having originated in Europe), fescue (which has four hundred to five hundred species), clover (which was a desired component of most lawns until herbicides were introduced in the 1950s), and a thirty-nine-by-seventy-eight-foot area of grass shorn five-thirty-seconds of an inch tall where a former University of Wisconsin tennis coach practices to maintain his status as the No. 1 over-eighty-five grass court tennis player in the world.

Koch explains there's no perfect grass for all situations. You can create a lawn that's beautiful and able to withstand tons of foot traffic, but it requires tons of care. You can develop a lawn that requires little mowing and maintenance, but it lacks aesthetics. "There's no silver—or green—bullet," he says.

There are those who argue about whether it's proper to have a lawn at all. They maintain that lawns replace natural ecosystems with unnatural ones that provide no food, wildlife habitat, or benefits; that lawn fertilizers and herbicides kill birds, animals, and insects, and, eventually, aquatic animals and plants. Those on the flip side point out that lawns help cool houses and cities, lower air-conditioning costs, and trap dust. They site evidence that most of the phosphorous in lakes comes from decomposing leaves, soil, animal waste, and natural sources, not fertilizers.

The anti-lawn community points out that Americans dump nine billion gallons of water a day onto their lawns—water that's been purified, chlorinated, and fluoridated to keep diseases out of our bellies and teeth.[2] And indeed, one-third of our treated water does wind up on lawns.[3] The pro-lawn faction notes that part of the issue can be resolved via education. People need to adjust their sprinklers so they're not watering concrete, fix their automatic sprinkler systems so they don't leak, and water early in the morning or late in the evening so water soaks in rather than evaporates in the midday sun.

Lawn opponents counterpunch that the gas-powered lawn equipment homeowners fire up for three billion hours a year generates massive amounts of carbon emissions, contributes to noise pollution, and causes personal injuries. One study shows that running a lawnmower for one hour produces the same amount of air pollution as driving a car 350 miles. Homeowners spill seventeen million gallons of gas while refueling their lawnmowers each year. That's more than the 1989 Exxon Valdez oil spill.[4] Fertilizers and pesticides require massive amounts of energy to produce, gas mowers generate noise levels of up to one hundred decibels— enough to damage hearing after just two hours—and to top it all off, there are six thousand visits to the emergency room each year in the United States because of lawnmower accidents. YET—new regulations are making equipment safer and more energy efficient. Cordless mowers are coming on strong. Mowing a lawn provides good exercise—you burn four hundred calories an hour with a push mower—and the peaceful repetition of lawn mowing can be, to some, like walking a Zen labyrinth.

Soil scientist Doug Soldat shows the crown and roots of a single grass plant, which has a life span of about eighteen months. Grass reproduces via underground rhizomes, since we usually mow it before it goes to seed.

This debate often remains civil, but not always. In 2017, Republican senator Rand Paul sustained six broken ribs and eventually had to have a lung removed after being tackled and thrown from his riding lawnmower by his neighbor—incidentally or coincidentally, a Democrat— who felt Paul was blowing lawn clippings onto his yard.[5]

What got us walking down this well-mown path to begin with? Some maintain it started when the fields and grasslands around castles and settlements were closely shorn so guards could spot invaders and predatory animals. Initially, the upkeep of a lawn required an army of scythe-swinging workers or a herd of grazing goats—something only the well-to-do could afford. Around 1670, André Le Nôtre, landscape gardener at Versailles, included *tapis vert* ("green carpet") as part of the design.[6] If Louis XIV could command one hundred noblemen to watch him dress, he could surely order that many servants to cut his grass.[7] Lawns remained the province of the rich for a century, but slowly they crept into the lives and yards of everyday people. A quintet of developments led to the booming popularity of lawns.

In 1830 Edwin Budding—inspired by a machine that would cut the "tufts and bobbly bits" from manufactured cloth—developed the first lawnmower, a rotating cylinder that pinched grass between blades and a stationary bar.[8] In short order, and with less hassle than maintaining a herd of goats, people of lesser means could afford a lawn. In 1920, gas-powered mowers were developed. When the cheaper, yet cruder, rotary mower was developed in the 1950s, nearly everyone could afford a well-kept lawn.

Second, landscape architects—most notably Frederick Law Olmsted, designer of Central Park and many early planned communities—made lawns an integral part of their designs. Olmsted "imagined a river of grass, flowing house to house as if the residents lived in a park."[9]

Third, lawn games exploded in popularity. Golf became wildly popular in the middle and late 1800s. Boccie ball, croquet, lawn tennis, and badminton required shorn grass, as did baseball, cricket, football, and soccer as they became popular.

Fourth, Orlando Scott, who had a "white-hot hatred of weeds," began selling grass seed and fertilizers directly to consumers in the early 1900s.[10] Advertising encouraged consumers to "think green" and to create "a lawn made for champions."

And last, but not least, planned communities began spreading like creeping charlie after World War II. Levittown led the charge. Each of the six thousand homes in the development came with a seeded lawn and a protective covenant that mandated they be mown once a week; those who failed might awaken to the sound of a maintenance worker mowing their lawn—and a bill taped to their door. There was even a political aspect to the new suburban lawn. Many of the homes were purchased by GIs—soldiers trained to be crisp, conformant, and obedient. In the watchful climate of the 1950s, those who had unkempt yards with crabgrass, well, they very well might have anticapitalist leanings. "No man who owns his own house and lot can be a communist," Bill Levitt explained. "He has too much to do."[11]

Today, about fifty million acres of turfgrass cover about 2 percent of the land surface in the United States. And people in the US and around the world work like hell to maintain their God's little acre.

"Mowing my lawn is my passion," explains Keith Smith of Britain, winner of the 2018 Creative Stripes Competition. "I don't drink or smoke, and my wife and kids are my world, and this keeps me fit and healthy."[12]

Brits are royalty of the well-mown hill when it comes to lawn care. It's home base for dozens of lawnmower racing clubs that sponsor "sprint" and "endurance" competitions; for those lacking a riding mower, there are "run behind" categories. A more covert competition takes place on the other side of the pond. A recent study shows that more than a third of American homeowners feel there's a friendly neighborhood rivalry over who has the best lawn.[13] But it's not just John and Jane Doe who love mowing. Actor and lawn-mowing devotee Richard Widmark wound up in the hospital after tangling with his mower. Widmark was so infatuated with mowing that after he finished off the forty-acre lawn of his Connecticut estate, he moved on to the neighboring estates of Walter Matthau and William Styron. After the mower accident, he didn't ask his doctors whether he would ever act again; he asked them whether he would ever mow again.[14]

You may be wondering whether there's a halfway point between the perfect lawn and one of dirt and thistle. There is.

Xeriscaping—using native plants accustomed to growing in native conditions—requires less irrigation and care. The California "Cash for Grass" turf-replacement rebate program, where eight million square feet of grass were replaced by drought-resistant plants, saves Los Angeles 250 million gallons of water a year.[15]

Pollinator and butterfly gardens—planted with the express purpose of providing bees and butterflies with necessary nutrients—are an environmentally healthy option.

Vegetable gardens—common up through World War II— have both culinary and social benefits. One proponent, who planted a front yard garden, noted, "People walking by— instead of just breezing past—stopped to talk to us. Some were just curious; others had tips and stories to share. We had more conversations with neighbors we'd never met in those few weekends than we've had in the entire 15 years living here."[16]

And if you love your lawn but don't want to use your time or gas to mow it, consider renting a sheep or goat. They can consume a quarter of their body weight in vegetation per day and are especially good at rooting out woody plants, poison ivy, buckthorn, thistle, and weeds. They fertilize too. The groundskeepers at the

Louvre in France, the Google campus in California, and Riverside Park in New York City use goats to "mow" some areas—just like the nobility of old.

This eco-friendly chèvre des Fossés (ditch goat) spends her day rooting out thistle and weeds in the Tuileries Garden near the Louvre in Paris.

HACKS & FACTS

Grass Does Look Greener on the Other Side

There is truth to the adage that the grass looks greener on the other side. When you're standing in your yard looking straight down, you see the dirt and bare spots between the blades of grass. But when you look over at your neighbor's yard at an angle, you see only green blades.

TREES

Stiletto Heels *and* Stumping *to* Survive

ONLY IN CARMEL, CALIFORNIA, CAN MAJESTIC TREES, HIGH HEELS, jagged sidewalks, a jab at urban life, and a dash of humor work their way into a city ordinance. Chapter 8.44 proclaims:

> It is recognized that much of the charm and appeal of the City to residents and visitors alike is due to its urban forest character, featuring the maintenance of Monterey pine, oak and other native trees or shrubs throughout the City.

The section continues:

> The maintenance of an urban forest throughout the City necessarily involves some informality in the lighting, location and surfacing of street and sidewalk areas, which in turn involves greater risk to those wearing high heeled shoes more adaptable to formal city life.

Thus:

> The wearing of shoes with heels which measure more
> than two inches in height and less than one square
> inch of bearing surface upon the public streets and
> sidewalks of the City is prohibited, without the
> wearer's first obtaining a permit for the wearing of
> such shoes.[1]

All tongue-in-cheek, yes.
But trees in Carmel are no
laughing matter. The city and
its residents give its urban forest priority over almost everything: roads are rerouted,
rooflines jog, sidewalks undu-
late, land developers rip their
hair out, Manolo Blahnik
shoe stores are sparse.

If you're remodeling or
building in Carmel, tamper-
ing with a tree, or even a tree
root, can be verboten. The
city rates each tree as "Significant," "Moderately Significant,"
or "Non-Significant." Taking
down a "Significant" tree on
private property can, as one

Carmel, California, reroutes sidewalks,
roads, rooflines, and everything else to
give priority to its urban forest.

landowner discovered, result in a $36,000 fine. Another contractor
was fined $18,000 for severing two large roots. Contractors need
to protect trees with wooden enclosures. When I ask Carmel city
forester Mike Branson about this, he muses, "Well, I guess it's sort

of strange to cut down trees to build a cage to protect other trees, but that's the way we do it in Carmel."

Of course, they do lots of strange things in Carmel. They elect Clint Eastwood mayor. They ban streetlights, house numbers, mailboxes, and—though repealed—ice cream. It's the only place I've visited where waiters glance at you suspiciously if you *don't* bring a dog with you onto the patio. But they sure get their trees right.

In 1989, President George H. W. Bush stood on a stage in Sioux Falls, South Dakota, and waxed (as poetically as H.W. could wax), "Trees can reduce the heat of a summer's day, quiet a highway's noise, feed the hungry, provide shelter from the wind and warmth in the winter. You see, the forests are the sanctuaries not only of wildlife, but also of the human spirit. And every tree is a compact between generations."[2]

Twenty-four years earlier, Ronald Reagan had waxed less poetically, "You know, a tree's a tree, how many more do you need to look at?"[3]

I'm voting with H.W. on this one.

If I were a tree stumping for office, I'd campaign on these points:

1. *I'll help you save energy.* Well-placed trees can significantly reduce your home heating and cooling costs. In urban areas, trees shade pavement, lower temperatures, and prolong the life of asphalt by up to 50 percent.[4]

2. *I'll provide jobs.* More than fifty million people worldwide—and two-and-a-half million in the United States—are employed in the forestry business; US trees add $600 billion annually to the world economy.[5]

3. ***I'm all about the environment.*** That formula for photosynthesis that was pounded into your head in eighth-grade biology ($6CO_2 + 6H_2O \rightarrow C_6H_{12}O_6 + 6O_2$) was more than a string of numbers and letters. It explains how I convert carbon dioxide (a greenhouse gas) and water into sugar and oxygen. In my first forty years of office, I'll absorb one ton of carbon dioxide, while supplying oxygen for four people a day.[6]

4. ***I'm good for the economy.*** I can add more than 15 percent to the resale value of your home. If you're a store owner, I'll increase foot traffic.

5. ***I'm there in times of catastrophe.*** I minimize erosion and flooding. My roots can uptake one hundred gallons of water per day.[7]

6. ***I won't mess with your social (or mental) security.*** I help reduce stress, lower blood pressure, even minimize road rage.[8] I have a calming effect on kids and adults with attention deficit hyperactivity disorder (ADHD).[9]

7. ***I can reduce your health-care costs.*** A Swedish study shows patients in rooms with "tree views" had shorter hospital stays, took fewer pain killers, and suffered fewer postsurgical complications than patients staying in rooms with "wall views."

8. ***I can quiet your life.*** Trees and shrubs, when densely planted, can reduce noise levels by five to fifteen decibels.[10]

9. ***I work for YOU.*** Each year, each of you uses the equivalent of twenty-five two-by-fours, six sheets of

plywood, and seven hundred pounds of paper and cardboard. My forest comrades and I provide every sliver of that.[11]

10. *I'm nonpartisan.* Regardless of your race, color, creed, gender preference, or political roots, I work tirelessly—24/7—for you.

So, ladies and gentlemen, THAT'S why I'm asking for your vote. Together we can make a tree-mendous difference!

But the life of an urban tree isn't necessarily made in the shade.

When I ask arborist Guy Carlson of SavATree about the chal-lenges of an urban tree, he tells me: "They're under enormous stress. Contrary to common belief, most tree roots are in the top twelve inches of soil—and so are streets, sidewalks, and lawns. A developer might have come in and scraped away the soil, removed all the nutrients, compacted the area, totally changed the natural environment, then spread around whatever soil was left and planted a tree—without even giving much thought to what *kind* of tree—and expected it to grow."

Carlson explains how decomposed leaves, which forest trees rely on for nutrients and natural mulch, are raked away in an urban setting. He makes a huge circle with his arms, explaining how a tree of that size often has to survive on a plot of land the size of his desk. "A tree can't defend itself a whole lot," Carlson muses. "Mostly what it can do is uptake water." Global warming hasn't helped matters. Like you and me, trees need a nightly "siesta" to recover from the day. When night temperatures stay high, trees don't get that recovery time—which leads to stress and risk of disease.

To get the water and nutrients it requires, a mature tree needs access to around fifteen hundred cubic feet of soil, the volumetric equivalent of a construction dumpster. Many boulevard trees get a tenth of that. But roots are ingenious, even mischievous; to get what they need, they'll crack open sewer lines, elbow through solid clay, push sidewalks out of their way. When all else fails, they'll simply limit their growth. Many cities are now installing engineered soil around trees and under sidewalks, giving trees access to more water and nutrients.

"They have to withstand road salt, snow, road construction, sidewalks, compaction, drought, and slamming car doors," Karl Mueller of the St. Paul Parks and Recreation Department says about the plight of urban trees. "A boulevard is sort of a hellish place for a tree." Some cities even designate trees as FHOs: fixed and hazardous objects.

But a tree's true adversary is much smaller than a car door— and when you combine this with trees all of the same species, the stage is set for disaster. I find this out the hard way.

I've arm-wrestled with depression for thirty years. Usually, it's a draw, but on a surprising number of days, I pin my opponent's wrist to the table. Yet occasionally the tables are turned. One gloomy day, while nursing a sore arm and recovering from a resounding defeat, I did what any reasonably depressed person with a sore limb would do—bought a three-acre walnut farm. I say "I" because, though it says "we" on the closing papers, it was a bit of a unilateral decision; the skid marks from my wife's shoes can still be seen in the closing office parking lot.

I/we bought the land midsummer, and our walnut trees were green and happy. But upon visiting "the farm" with friends five weeks later, we found the trees standing naked and afraid. No words were uttered, but all thought, "Spike bought three acres of terminally ill walnut trees." Upon doing late due diligence, I discovered there were indeed a litany of diseases, bugs, and blights that could wipe out our entire lot of walnut trees. I had purchased a monoculture.

Over the past 125 years, North America's urban and rural forests have been repeatedly plundered by disease, primarily because we, or Mother Nature, have established *monocultures*—vast forests and avenues of single species trees. One horticulturalist refers to this as "S5" (Simple Single Species Syndrome Sickness).[12] When a single species dominates, a single invader no larger than your fingernail can decimate forests, parks, and boulevards—and three-acre walnut farms.

Having rightfully earned their title King of the Forest, chestnut trees were the first to fall in the United States. At the dawn of the twentieth century, they were the most abundant and beloved tree in the country. Rural folks loved them for their beauty and the economic bounty they provided; the rot-resistant lumber was used for houses, barns, fences, and furniture. Urbanites valued them for their broad shade canopies. Kids loved them for their climbability. And everyone cherished them for their "roasting on an open fire" hominess. But in 1904, Hermann Merkel, chief forester of the Bronx Zoological Park, realized the chestnut trees in his kingdom were in crisis.[13]

The enemy was a fungus that had been introduced via Japanese nursery stock. The foreign fungus could enter even the smallest wound and then work its way beneath the bark to create pustules. Each pustule contained millions of spores, which could be spread by insects, birds, squirrels, and wind. Spraying, pruning, and cutting buffer zones all failed. Within a few years, a blizzard of spores killed every chestnut tree in New York City. Nationwide, four billion chestnut trees—and the economies and traditions that had developed around them—died. Squirrel, bird, cougar, and moth populations plummeted. Some labeled it "America's first and worst natural eco-disaster."[14] "The American chestnut tree survived all adversaries for 40 million years," the American Chestnut Foundation laments, "then disappeared within 40."[15]

Two more nationwide tree epidemics—elm and ash—struck during the ensuing century. University of Minnesota forestry professor Gary Johnson tells me that in many cities, 90 percent of the boulevard, park, and urban trees were elms. "They were tough as nails and grew fast," he says. "They could thrive in wet, dry, or compacted soil. Snow, ice, and air pollutants didn't bother them. Even utility companies loved them since they grew in a 'vase shape' around power lines." One observer described elms as "a fitting ornament to stand by the stateliest mansion or the humblest farmhouse." Another championed the elm "as rugged as a weed. It can live in almost any filth of smoke and soot and noxious fumes."[16] But they couldn't tolerate Dutch elm disease. Introduced to North America in 1928, hidden in a shipment of veneers from the Netherlands, elm bark beetles and the fungus they spread eventually wiped out three-quarters of America's urban elms. France lost 90 percent of its elms, Toronto 80 percent, the United Kingdom twenty-five million. It was indeed a "Nightmare on Elm Street."

Ash trees—similar to elms in toughness and rapid growth— often replaced elms and soon constituted up to half the trees planted in many urban and suburban areas.[17] But they too were discovered to be susceptible to monocultural plunder. The damage occurs when emerald ash borer (EAB) larvae tunnel under the bark, creating S-shaped trenches that block the flow of nutrients. An EAB infestation can kill a tree in one year. First discovered in Michigan in 2002, the problem spread nationwide in less than a decade. On its own, the EAB can travel only about one mile a year, but when hitching a ride on firewood it can travel hundreds of miles in a single day.[18] Caught in its early stages, the EAB can be slowed or halted with insecticide injections and other measures, but the battle is costly and often a losing one.

While similar to elms in life, ash trees differ drastically—and dangerously—in death. "An elm tree can die, become bare-boned, and stand for ten or twenty years without losing limbs, giving cities plenty of time to take them down," Johnson explains. "But ashes become an immediate hazard. They can start shedding limbs and breaking apart—even explode—before they're dead. Structures, vehicles, and people on streets, parks, schoolyards, and golf courses become unwitting targets. Their removal has created a huge financial burden for lots of cities."

A host of regional tree diseases have affected other areas, especially in the exurbs where people buy a few acres of natural forest, build their dream home, then impose the stresses of city life onto the more laid-back forest trees. In the 1940s, oak wilt began taking its toll in the Northeast and Midwest. "They're the developments we call Prior Oaks," Johnson explains with a half grin. Lethal yellowing disease is ravaging palms in the southeastern United States. Pine beetle blight has killed entire forests in the western states, providing standing kindling for lightning strikes and electrical fires. Johnson points to the recent spate of fires in California. "It's a fait accompli," he says. "You're surely not going to prevent lightning strikes. They're going to burn."

Arborists have finally gotten the message, and the message is diversity. Many urban foresters now follow the "15-10-5" rule, which recommends planting no more than 15 percent of trees from one family, 10 percent from one genera, and 5 percent from one species. "There isn't a perfect urban tree," Mueller explains. "You never know what [disease] is on the horizon. The only answer is to plant a mixture."

Trees Make Cents

Five strategically planted trees can cut your air-conditioning costs by 25 percent and your heating costs by 10 percent. California's nine million street trees provide enough cooling and electrical savings to air-condition half a million homes annually.[19]

What grows up must come down—and it's SavATree Carlson's job to make sure the second half of the proverb gets done safely. I spend a morning with one of his crews. The scene resembles more of a SWAT team raid than a horticultural event. There are boom trucks, skid loaders, cranes, and wood chippers working in a well-choreographed hydraulic ballet.

"There go some of my babies," explains Denis Heuer, who's lived on the property for sixty years and has planted most of the trees. It's early winter—a time when trimmers prefer to work since there's less foliage to work around, less chance of spreading disease, and firmer ground for supporting vehicles. Each member of the team has a specialty. One duo operates the aerial bucket, removing branches from one side of the tree on the way up and from the other side on the way down. Between the whine of the chainsaw and the roar of the truck, verbal communication is impossible. Head nods and experience do the talking.

Meanwhile, another crew is in the backyard removing trees the old-fashioned way: with chainsaws, ropes, and brawn. I watch one fellow don climbing harness, climbing loop, spiked boots, hard hat, chaps, and gloves and then tie a sixteen-inch chainsaw to a four-foot rope dangling from his belt. He scuttles up the tree and, every few feet, adjusts his climbing belt and safety rope. He's squirrel-like in his ability. While working, he locks himself into a position that only he sees as secure, grabs the chainsaw, starts it with a single pull, gives the branch a quick one-handed undercut, and then grabs the branch with his left hand and finishes the cut. I've attended three performances of Cirque du Soleil yet have never witnessed such a deft combination of balance, strength, flexibility, intuition, and balls.

The trees in Heuer's backyard are being pruned and thinned for a variety of reasons: some are being removed because of oak wilt disease, some because overhanging limbs are poised to crash through the roof, some are shading a garden area, others are becoming spindly from growing too close to one another. Selective thinning and pruning—though sometimes painful—is one of the keys to keeping trees and landscapes healthy. "I can't count the number of times I've hugged someone or watched them cry as we removed a dead or diseased tree from their backyard," explained one arborist. "It's like they're burying an old friend."

We return to Carmel and to the tale of one group that tried to take the tree-thinning notion way too far. In 2005, Clint Eastwood, Arnold Palmer, and their investment group pushed to build an eighteen-hole golf course and development in the nearby Pebble Beach area. An ad campaign designed to gain public support for the project touted a "save the forest" theme. The group declared they would donate four hundred acres of adjacent forest to be preserved in perpetuity as quid pro quo. What they didn't mention was that the development would involve cutting down eighteen hundred trees—fifteen hundred of them iconic Monterey pines.[20]

Maria Sutherland, president of the Friends of Carmel Forest, tells me: "I never set out to be a tree advocate, but this was too much. I told a friend about it and she said, 'Well, there's not much you can do about that.'" But Sutherland *did* do something about that. She teamed up with Mark Massara, "the surfing attorney" and then-director of the Sierra Club's California coastal campaign. Sutherland and other volunteers created a short documentary outlining the impact of the project. It aired on the local tourist channel. This, with the groundswell support of other groups and citizens, put an eventual halt to Eastwood's project. It did not make his day.

Maintaining Carmel's urban forest for a century has not been without other snags. Sutherland mentions one past city official who had been in power sixteen years and had slashed the forestry budget. "He was, at best, tree neutral," Sutherland says. "He had an artificial lawn." She also discusses the Friends group having planted a memorial tree only to have it poisoned or chopped down three times. The suspects? Residents of a nearby home where the tree might have impinged on their ocean view.

Still, the majority of Carmel residents remain arboreally vigilant. During public commentary at a city council meeting I attended, six of the dozen people who spoke addressed tree issues. One couple complained that the city wasn't caring enough for trees. As an example, they said they had had to carry five-gallon buckets of water two blocks to rescue a dying sapling. One of the main agenda items revolved around the location of a new driveway and the number of trees it might affect. The debate went on for ninety minutes. The variance failed.

If you want to have a long-lasting effect in your neighborhood, plant a tree. If you do, there will then be three trillion and one trees on the planet.[21] This may seem like a surfeit, but we're losing trees at a rate of ten billion a year. Because there are sixty thousand species to select from, you'll have no shortage of planting options.[22]

Given these huge numbers, you may feel your one tree is a drop in the bucket, but the world is made of drops in a bucket. Wangari Maathai, a woman who grew up in the rural village of Ihithe in Kenya, started a tree-planting movement that put thirty million drops in the bucket.

When you plant a tree, you plant a timeline. Nothing in nature keeps pace with the human life cycle as well as a tree. Rocks last eons, flowers last weeks, birds come and go, but trees remain rooted. They grow in tandem with us. The scrawny, uncertain "Charlie Brown" sapling my family planted fifteen years ago has become a living diary. The first year it slept, the second it crept, and the third it leapt. The fourth year it welcomed a grandchild into the world, and the twelfth it welcomed her into its branches. It's served as a backdrop for wedding photos, anniversaries, and birthdays. It's opened its branches to birds, swings, and tangled kites. Like a growth chart etched on the edge of a door, it's a reminder of times of both growth and stagnation. It's a good companion for sitting and reminiscing.

WALK THE WALK

Three Simple Things You Can Do to Make Trees Happy

Trees are so big and seemingly self-sufficient, most of us just let them do their own thing. But forestry professor Gary Johnson explains that every tree can live longer, grow stronger, and work harder with a little help.

1. *Water them.* "Most people think, 'Well, it's going to rain this weekend' or 'It's done fine on its own so far,'" Johnson says, "but water is the magic bullet. It's the single most important thing you can do to help a tree— especially a recently planted one." One rule of thumb says, in dry conditions, water a mature tree five minutes for each inch in diameter, once or twice a month. For new trees, give them two gallons of water for each inch in trunk diameter, twice a week.

2. *Fertilize them.* "Mixing a good organic fertilizer, like composted manure, into the soil provides nutrients and helps loosen up and condition the soil," Johnson says. "If you want a tree to grow faster, this will give it a boost."

3. *Mulch around them.* "The compost from your local composting site is black gold," Johnson enthuses. "Spread it three or four feet out from the base to prevent compaction, help retain moisture, and minimize lawnmower bumps."

SQUIRRELS

The Gnawing Truth

I'M SITTING IN JOHN MORIARTY'S OFFICE. HE HAS A SIX-FOOT bull snake slithering around atop his filing cabinet, a license plate imprinted with TOR2GA, and a badge emblazoned "Sr. Wildlife Manager, Three Rivers Park District." Which makes me think I've found the right guy to talk to about squirrel behavior.

When I ask him about the squirrel's uncanny ability to thrive in so many environs—urban parks, suburban backyards, wilderness forests—he explains, "They flourish because they—like humans—are 'generalists.' Squirrels can live in treetop nests, hollowed-out tree trunks, or your attic. They'll eat acorns, bugs, tree bark, tomatoes, or the tulip bulbs you planted this morning. Some animals—like the Regal Fritillary caterpillar we're trying to introduce into the park that will only eat violets—need very specific conditions. But squirrels can live anywhere." And they do . . . almost.

Squirrels evolved around forty million years ago and—like horses, dandelions, and cockroaches—originated in North America, then spread via ancient land bridges. They've since evolved into 278 species and can be found on most parts of the planet except for Australia, Greenland, and the Galapagos Islands. Worldwide, squirrels range in size from the three-inch African pygmy squirrel to the forty-inch Laotian giant flying squirrel. They've been found at elevations as high as fifteen thousand feet and can swim, though as we'll see, not particularly well. One species of flying squirrel can glide more than three hundred feet, and one common gray squirrel, in an unsanctioned standing broad jump competition, was able to leap 8.2 feet.[1] Squirrels belong to the Rodentia biological order of mammals but are members of the Sciuridae family, meaning they are not, as some claim, "simply rats with good PR agents." Not all squirrels fit the standard squirrel profile; prairie dogs, chipmunks, and marmots are all biologically classified as squirrels.

With 106 squirrel species, southern Asia is considered the "Squirrel Capital of the World." Europe has a measly seven species—and would prefer to have six since the American gray squirrel is considered an invasive pest. North America has 66 species, with a density of about 1.5 squirrels per acre—making your block home to 8.1 squirrels. And since the range of most squirrels is only a few acres, chances are the same old squirrels you see day after day in your backyard are exactly that.

Given the diversity, we'll narrow the field by focusing on fox, red, and eastern gray squirrels, which are common in urban and suburban settings and are the squirrels you'll most likely encounter on your walk around the block.

The squirrel is a fine-tuned biological machine. The bushy tail does more than make them cuter than their universally reviled cousin, the rat. The tail is an indispensable balancing aid—a sort of Wallenda-esque balancing pole—used to maintain equilibrium as they leap from branch to branch or run along overhead wires. The tail also is a heating and cooling system; the complex circulatory system within it serves as a heat exchange mechanism that keeps the squirrels warm in winter and cool in summer. In extreme cold, squirrels use their tails as blankets, in extreme heat they use them as parasols, and in times of extreme danger, as alarm systems.

The flexible anatomy of a squirrel, which allows them to rotate their back feet 180 degrees when they descend trees headfirst, puts contortionists to shame. One researcher equates this ability to a human standing on tiptoe, rotating the ankles so the soles of the feet face each other, then continuing to rotate the feet until the soles point forward. This flexibility is a survival mechanism: it allows squirrels to circle trees as they ascend or descend, giving them a 360-degree view of the world so they can watch for and evade predators.

Swiss Army knife–like in their versatility, a squirrel's incisor teeth can open nuts, chew holes in plastic storage containers, and gnaw through wood doors with ease. Since the four front incisors grow six inches a year, squirrels—like other rodents—must constantly eat or gnaw to keep the length in check and keep them sharp. What appears to be mischievous chewing behavior to us is survival for a squirrel; if the incisors grow too long, the squirrel can no longer chew and will die.

Squirrels are primarily loners. During mating season, the males are the love-'em-and-leave-'em type. Females are nurturing while raising their young, but when the young leave the nest, it's pretty much every squirrel for itself.

People either love squirrels or hate them. On the love side, you'll find the citizens of Longview, Washington. In response to an inordinate number of squirrel fatalities on Olympic Parkway, Amos Peters built an overpass in 1963. Nicknamed "Nutty Narrows," it consisted of a sixty-foot length of retired fire hose in the form of a suspension bridge. It's so iconic, it's been placed on the National Register of Historic Places alongside the Golden Gate and Brooklyn Bridges. Five more squirrel bridges have since been built. The town mascot is Sandy B. McNutt the Squirrel, and the town celebrates its Sciuridae-al passion with its annual Squirrel Fest.

Those living in Olney, Illinois—the city with the highest population of white squirrels in the United States—are equally squirrel-centric. Squirrels have the right-of-way on streets, and reckless drivers are fined $750 in the event of a fatality. Cats aren't allowed to freely roam, and dogs must be kept on a leash.

The citizens of Tyler, Texas, were so enamored of Shorty the Squirrel that for fifteen years, they put up with his daily panhandling outside the Smith County courthouse. Citizens provided free medical care and a special squirrel-destrian crossing. The legend of Shorty grew to nutty proportions when Paul Harvey—himself a member of the Squirrel Lovers Club—featured him on his radio show.

Kelly Foxton, a former military pin-up queen turned country-western singer, has spent $500,000 on custom-made outfits (including fur coats) for the half-dozen pet squirrels—all named Sugar Bush—she's kept over the years. "People think I'm nuts and I don't care because I'm not," Foxton says. "I have a 156 IQ."[2]

But perhaps no one has loved a squirrel more than Tayfun Demir. Demir, a software engineer from Istanbul, drove seventeen hours to rescue a squirrel, which he eventually christened Karamel. The injured squirrel had both arms broken in a close encounter with a hunter's trap. Demir paid $500 for two surgeries—the first to set the broken arms in casts, and the second to have the arms amputated when the first procedure failed. Demir had a prosthesis built and fitted. It resembled the front half of a roller skate. Today, Karamel leads a near normal life, fending for herself in Demir's backyard. "Karamel is spiritually the strongest squirrel I have ever seen," Demir told me.

If I'd found Karamel in my backyard, I could have taken her to the Wildlife Rehabilitation Center of Minnesota a few miles away. Last year the center admitted 2,160 gray squirrels—along with

Prosthesis developed by Tayfun Demir of Istanbul for Karamel, a young squirrel that lost both arms in a close encounter with a hunting trap.

2,839 rabbits, 603 robins, and about 8,000 other patients of nearly two hundred species. Her chances of survival would have been about fifty-fifty. The Humane Society website has contact information for the thirty-nine other states that have similar centers.[3]

But all is not fur coats and celebrity status for squirrels. Bill Adler, author of *Outwitting Squirrels: 101 Cunning Stratagems to Reduce Dramatically the Egregious Misappropriation of Seed from Your Birdfeeder by Squirrels*, maintains, "If the creativity and energy that bird feeders put into thwarting squirrels were directed toward world peace or eliminating traffic jams, we'd have no more earthly problems."[4]

It *is* ironic that bird lovers are often the most avowed squirrel haters. Squirrels lust after sunflower and other seeds, and because 40 percent of North American households maintain bird feeders, suburban squirrels rarely need to range far for food. Squirrel deterrents abound. People attempt to protect bird feeders with baffles, barbed wire, and Bengay. One suggested deterrent is covering bird feeder poles with Crisco, or embedding dog hair or jalapeño peppers.

"Squirrels are more of a nuisance than a problem," Moriarty says. "They don't cause significant damage. They don't spread disease. They don't have parasites. Deer can decimate an entire garden in an hour and moles can wreck your yard, but you can control most squirrel problems. If they're chewing through your plastic bird food container, use a metal one. If they're running all over your roof, cut back the branches."

They can inflict the most serious damage when they breach the perimeter of a home and set up camp in an attic. To satisfy their biological need to gnaw, they'll chew ceiling joists, PVC pipe, electrical wires, and plaster. Seat-of-the-pants solutions for attic evictions include playing loud music, installing strobe lights, and scattering cat feces.

When asked about more tried-and-true solutions, Moriarty simply smiles and says, "Carpentry"—which is the approach most critter-control specialists take. First comes squirrel eviction. The most common solution is to position a one-way door over the entry hole, then seal the opening once the squirrel exits. For squirrels refusing to leave peacefully, live trapping and relocating is often the next step. "Some people think relocating a squirrel is the humane thing," Moriarty says, "but it's like someone picking you up and plunking you down in a strange neighborhood then saying, 'Okay, this is your new home.' Squirrels aren't used to the new territory, there's a hierarchy, they get beat up or try to find their way home and they don't make it." In fact, 97 percent of all squirrels relocated from the burbs to forest don't survive.

But what have squirrels done for *you* lately? Well, they're some of nature's most prolific tree farmers and sowers of seeds. Maple seeds have helicopter wings that can fly two miles in strong wind, and the sandbox tree has exploding pods that can launch seeds the length of a football field.[5] But nut trees have only the thunk and roll of gravity—and squirrels—for their distribution system. The

20 percent of nuts that squirrels bury and never return for have been given a proper planting. The amnesiac nut farmer has buried them the right depth, cultivated the ground around them, and perhaps left a dab of fertilizer to nourish them.

Squirrels that fancy a good mushroom also help distribute spores via their feces. Squirrels help aerate the soil with their digging, eat beetles and grubs that can infest trees and lawns, and even help keep some plants pruned. That's what they've done for you lately.

Some people estimate that squirrels spend 75 percent of their waking hours locating, acquiring, hiding, and consuming food. They're omnivorous and, while preferring plants, will in a pinch eat bugs, baby birds, and eggs. Robert Lishak, a professor at Auburn University who's studied squirrels for forty years, explains to me squirrels' uncanny ability to locate food: "They queue in on other species' alarm calls. If you put food out, chances are birds are going to be the first to spot it. If squirrels hear a crow calling or a blue jay screeching, they're quickly on the scene to investigate."

HACKS & FACTS

Big Eaters

Squirrels consume roughly one-and-a-half pounds of food per week—an amount equal to their body weight and the human equivalent of three grocery shopping carts of food.

Red squirrels are known to collect mushrooms and dry them before stashing them. Hickory nuts—with a fat content of 30 percent and shells just the right hardness on the nut Mohs' scale for sharpening teeth—are their beef bourguignonne.

The 1988 BBC documentary *Daylight Robbery!* examined how many obstacles a squirrel was willing to learn how to negotiate to access food. The course consisted of spinning baffles, wire tight-ropes, PVC pipe tunnels, barrel-rolling disks, death-defying platform-to-platform leaps, and a plunger push—fourteen obstacles in all. One squirrel completed the course twenty times in a single day.

The squirrel's instinct to hoard food is legendary. In a single hour, a squirrel can bury twenty-five nuts or put up a cache of pinecones large enough to fill an engine compartment. Squirrels squirrel away food in one of two ways: fox and gray squirrels employ *scatter hoarding*, while Douglas and red squirrels use *larder hoarding*.

Scatter-hoarding squirrels—clearly not readers of *The Life-Changing Magic of Tidying Up*—bury or store individual or small stashes of nuts over a wide area. Some researchers maintain they can "remember" the location of a buried nut for only twenty minutes but can compensate for their forgetfulness with a keen sense of smell. Scatter hoarding is not without purpose; not all of a squirrel's nuts are in one basket in the event of a natural disaster or nut raid. And, if a squirrel does move its nest, chances are a hoard will be nearby for sustenance.

Larder-hoarding squirrels *do* put all their eggs in one basket—a basket called a *midden*. The advantages are (1) there's less to remember, and (2) they wind up with a near-guaranteed access to plentiful food throughout the year. The disadvantage is that the midden must be defended, which requires time and energy and a certain feistiness. If you encounter a two-foot-tall pile of pinecones at the base of a tree or other place, you've discovered a midden—and that stash may have fifteen thousand cones or more.

Food hoarding does not prevent all problems. A prolific acorn crop on the East Coast in 1967—a mast year—led to the 1968 Great Squirrel Migration, as squirrels searched out more places to hide their bounty. Many sought greener pastures on the other side of bodies of water. An estimated one hundred thousand drowned squirrels were subsequently pulled from one upper New York state reservoir.[6]

Squirrels are prolific. They reach sexual maturity at ten months and, though receptive for only eight hours at a crack, can breed twice a year. Each litter can contain as many as four babies or *kits*.[7] But they *have* to be prolific. The mortality rate of squirrels during their first year is as high as 75 percent. Kits are nest-bound for the first eight weeks, easy prey for marauding cats, various birds, and inclement weather. If something happens to their caretaker, baby squirrels don't stand a chance. While some squirrels can live more than a decade, the average life of a squirrel in the wild is only five months, though squirrels that make it through the first year have an ever-increasing rate of survival.[8]

A variety of factors keeps the squirrel population in check. Electricity is one. A squirrel's body is just the right length to act as a jumper cable between the "hot" and "ground" contacts of some electrical wires and transformers, places where squirrels often visit to stash food or, if pole-mounted, begin a high-wire journey. Some people estimate the price tag of squirrel-induced electrical damage to be $2 billion per year in the US alone. Squirrel electrical shorts have shut down trading on the Nasdaq stock exchange not once, but twice. And "squirrel-ages" are so frequent in one neighborhood that one disgruntled resident proclaimed, "We don't even bother setting the clock on the microwave anymore."

When I was an editor at *Family Handyman* magazine, a tip for keeping squirrels out of the attic was submitted for the popular Handy Hints department. The advice was to cut off the female end of an extension cord, wrap the wire ends around two nails, place the nails "squirrel width" apart near the entrance hole, then plug in the cord. Invading squirrels that brushed between the two nails would complete both the electrical circuit and, presumably, their life cycle. A few weeks later another envelope arrived bearing large letters, "OPEN IMMEDIATELY." Within we found a six-word note: "DON'T RUN TIP. HOUSE BURNED DOWN."

Squirrel hunting is still alive and well. In Minnesota more than a quarter million squirrels are "harvested" a year; in Louisiana, it's three times that number.[9] The meat tastes "sweeter than chicken and richer than rabbit," according to one source.

Holley, New York, is home to the annual Squirrel Slam, a hunting contest that draws up to three hundred participants—and an equal number of protesters. One hunter maintains it's no different from the hundreds of fishing contests held each year. "This is a way of life up here. It's really no different than a fishing derby. You need a license and it has to get weighed."[10]

One avid squirrel hunter, Ted Nugent, "the Motor City Madman," known as widely for his conservative views as his guitar riffs, lays claim to the world's distance record for picking off a squirrel using a bow and arrow at 150 yards.

At least two British butcher shops can't keep squirrel in stock because it's so popular; the proprietor of Ridley's Fish and Game sells as many as sixty a day. But food critic Jay Rayner feels squirrel isn't destined to become a regular part of most diets. "People may say they are buying it because it's green and environmentally friendly," he explains. "But really they're doing it out of curiosity and because of the novelty value. If they can say, 'Darling, tonight we're having squirrel,' then that takes care of the first thirty minutes of any dinner conversation."[11]

And there are cars. The first instinct of a squirrel in danger is to freeze; the second is to run in a zigzag pattern to throw off predators. Though able to run twenty miles per hour, neither defense works well when dodging traffic moving forty-five.

The journey of the urban squirrel has not been an easy one. As city populations flourished, native squirrel populations diminished. Their chances of survival were slim since most early cities were devoid of parks and squirrels were also a source of food.[12]

In the mid-1800s, spurred by the design of expansive parks, squirrels were introduced to city parks. The first importations were a boondoggle. Placing squirrels in barren parks was like stocking trout in a swimming pool—plenty of room, no food. But as parks, trees, and the experience of urban foresters grew, so did the squirrel population. People began viewing squirrels as a positive force. Etienne Benson of the University of Pennsylvania argues that squirrels were an integral part of bringing nature into the city to boost people's health, to provide leisure activities for workers unable to escape the city, and even to serve as moral educators.[13] Feeding squirrels encouraged humane behavior, and squirrels, in turn, taught humans the benefits of stashing away provisions for meager times.

Today, urban squirrels thrive. Lafayette Square, a seven-acre park across from the White House, may support the highest density of squirrels anywhere in the United States. Three squirrel purges in the mid-1980s, spearheaded by a squirrel whisperer, led to the relocation of nearly a hundred squirrels, yet the squirrel population continues to thrive thanks to the abundance of natural and tourist-tossed nuts.

We're quick to assign human characteristics to squirrels. They're *cute.* With their bushy tails, chubby cheeks, and ability to "eat with their hands," they've endeared themselves to park goers, nature lovers, and children's book authors throughout history. They're *playful.* Though most tail-chasing and tree-circling are squirrel foreplay, hierarchy reinforcement, and communication, they do look like they're having fun.

But are they *intelligent?* When I pose this question to Lishak, he explains: "They learn via the typical learning paradigms—operant and Pavlovian conditioning, habituation, and so on. Most of their behaviors are instinctive, but still, learning factors in—because every behavior is both nature and nurture. When it comes to problem solving, squirrels are unique. If you have a dog tethered to a twenty-foot leash and the leash gets hung up on a stump ten feet away while he's trying to get at his food, he'll strain at the leash and bark and get himself tangled up—but he can't figure out another way to get to the food, because he lives in a two-dimensional world. But squirrels live in a three-dimensional world. They climb and can see depth, up, down, left, right, to and fro. And this makes them incredible spatial problem solvers. They'll actually move away from the food source to find another route." Which—in the end—makes them look intelligent.

They do have communications skills. Lishak used spectrographs, model cats, and real cats (restrained by a tether) to study squirrel communication. "If a cat is just loping along at an even pace, squirrels ignore it," explains Lishak. "Stalking—starting, stopping—sets off alarm signals. If the cat makes eye contact, it sets them off in a New York minute."[14]

An initial *kuk, kuk, kuk* warns other squirrels that a predator (usually a cat) is lurking and also informs the predator it has been spotted. "The squirrel is basically saying, 'Hey, you've been spotted. Your chances of successful predation are slim to none; don't waste your time slinking around in the bushes,'" Lishak tells me. "And, sure enough, the cat will usually walk away realizing there's no way to be successful."

If the predator continues to approach in a threatening manner, the frequency of the *kuks* increases, the squirrel faces the direction of the predator, and rapid tail flagging ensues. "If you're another squirrel in the woodlot, just looking at the silhouette of the squirrel

Mechanical cat used by Robert Lishak to study squirrel communication. A rapid series of *kuks* send a warning to other squirrels—and informs the cat, "You've been spotted. You're busted. Don't waste your time lurking around in the bushes."

and listening to the *kuks* tells you where the predator is and how fast they're approaching," Lishak explains. *Quaaaa* signals that a predator is still in view but is moving away. A less intense *quaaaa-moan* indicates the predator has (probably) "left the building."

A muffled *muk-muk* means two things: from a nesting female, it signals a need for food; from a male, it indicates a need for a Tinder date. And tooth-chattering is one squirrel's way of telling another, "The next thing I'm going to do is growl . . . and then I'm going to bite you," Lishak says.

Squirrels, all 1.12 billion of them in the United States, are in our hearts and minds . . . and attics . . . and bird feeders. You don't have to love 'em, but you sure as hell can't leave 'em—because they won't leave you. But perhaps you'll be more understanding of a squirrel's motives next time you find him or her gnawing on your antique wicker porch furniture or trying to wrestle an avocado seed out of your trash bin.

SNOW

The Fate *of* My
3,358 Inches

AS A LIFELONG MINNESOTA RESIDENT, I'VE SLOGGED, SHOVELED, and skidded my way through sixty-six winters' worth of snow. Out of curiosity, I plowed through the records to tabulate my personal ISE (Inches of Snow Endured.) Total? 3,358 inches . . . a few flakes shy of 280 feet—enough snow to bury a thirty-story building. Many might ask, "Why would anyone live in such a god-forsaken place?" My question is "What the hell happened to my 280 feet of snow?"

To find the answer, I needed to find someone in charge of plowing 110 miles of roads and 150 cul-de-sacs, someone who knows the ups and downs of driving a fifteen-ton snowplow through a blizzard with morons whirling around them, someone who knows how to govern a 1.3-million-pound pile of salt. I needed to find Joe Keding, street supervisor for the city of Shoreview, Minnesota. After spending a morning with Keding, it became clear that plowing snow involves more than blades of steel. It involves nerves of steel, meteorology, cartography, ecology, public relations, and a knack for fortune-telling.

I begin by asking Keding how he handles a typical snowfall. "No such thing," he responds. He opens a book containing maps of the city and points to the roads marked in red; if it only snows an inch or two, his crews clear only the "red mains." More than that and they go to "full plow" and tackle the "gray roads" too. It takes about five hours—and costs the city about $10,000—for his seven drivers to complete a full plow for this city of about thirty thousand.

The best-case scenario is a snowfall beginning late at night and quitting by early morning, so his crew can begin plowing at 2 a.m., which gives them enough time to clear the roads before rush hour. The worst-case scenario is a heavy snowfall at 5 or 6 in the morning, when plows spend precious minutes, sometimes hours, idling in rush-hour traffic and dodging vehicles. By law, his crew can work shifts of up to fourteen hours long, and in heavy snowstorms, they do.

"Believe it or not, we do our best to *not* make people's lives miserable," Keding continues. "All we really want to do is clear snow." But the inverse does not hold true. People leave trashcans and cars on the street, which makes it impossible to clear snow curb to curb. Lead-footed drivers follow too close and pass on the right. Homeowners complain about windrows of snow blocking their driveways, without understanding that plow drivers don't have the time or equipment to "unblock" them. (If you want your plow driver to love you—and want to avoid redigging out the end of your driveway—clear a strip on the "approach side" of your driveway for the plow to dump its accumulated snow.)

Todd Stevens, an engineer with the Minnesota Department of Transportation (MnDOT), explains to me: "Plows rarely travel the posted speed; at faster than thirty miles per hour any sand and salt they're spreading just bounce off the road. Yet people will just barrel into a cloud of snow not thinking about what's causing it. Plows have flashing red lights and lots of other warning lights. We're dumbfounded how people can run into a plow." Yet it happens all the time; 10 percent of MnDOT's eight hundred plows were involved in some form of calamity in a recent year.

"Our drivers are dedicated," Stevens continues. "Our department will tell people to stay off the roads because conditions are terrible, then turn around and tell our drivers to get out there and plow. They get woken up at 2 in the morning and pull twelve-hour shifts, they miss kids' sporting events and Christmas dinners. They're underappreciated."

Many drivers don't understand that if a plow is going to completely clear a road, the plow blade has to be over the centerline in one or both directions. "Being stuck behind a plow is always better than passing one," explained one veteran snowplow driver. "I don't know how many times I've had someone pass me, then I pass them in the ditch a minute later. I can usually pick 'em out; they're the ones with a cell phone in one hand and a burger in the other."

Even when caution is exercised, a fifteen-ton orange projectile, fronted by a ten-foot blade flinging snow at thirty miles per hour, is destined to inflict damage. Last year, Keding and crew repaired sixty torn-up boulevards and compensated homeowners—$50 a pop—for an equal number of coldcocked mailboxes. Labor-wise, homeowners are responsible for replacing their mailboxes, but Shoreview is just small enough and Keding's heart just big enough that, if his crew takes out a little old lady's mailbox in January, they'll probably be out there in April with a posthole digger.

"It's a very nerdish thing to do, but most cities have a snowplow ordinance and people should know it," Keding says, adding a bonus tip: "If you're shopping for a house, never buy the first house on the right in a cul-de-sac. When drivers turn that corner with a full plow of snow, it's gonna end up in your driveway."

The nearby metropolis of Minneapolis follows the same basic game plan—times ten. Each time it snows, the city plows the equivalent of clearing a path from Minneapolis to Seattle and back.[1]

Throw in 3,700 alleys, 57 miles of parkway, and 250 bridge sidewalks and it's easy to consume the annual budget of $13 million. Even in a snow-starved winter, the budget is exhausted by the cost of preparedness—personnel, equipment, and materials that must be at the ready, snow or not.

In urban areas, when there's no place left to push the snow, truck-mounted snow throwers are often employed. But when there's no place to throw, removal is the only option—an expensive one requiring front-end loaders, dump trucks, and traffic control. In the past, snow—salt, hubcaps, oil slicks, and all—was often dumped into nearby bodies of water, until the EPA said, "No more."[2] Cities needed to create dumping grounds. One winter, Minneapolis generated a 180-foot-tall pile of snow.

One way to accelerate Mother Nature's melting process is with "snow melters"—dumpster-size tanks of hot-water baths or sprays into which snow is dumped. The meltwater is filtered and then piped into holding ponds, storm sewers, or sewage treatment plants. Melters can burn sixty gallons of fuel an hour, but compared with trucking costs, they are a fossil-fuel saver. New York City has thirty-six melters, each one capable of melting 60 to 130 tons of snow per hour.

Holland, Michigan, averages seventy-six inches of snow per winter—and plows almost none of it. There, beneath the pavement, lie 168 miles of tubing that circulates heated waste water from the local power plant. The tubing can melt one inch of falling snow per hour. The initial installation costs are high, but once the coils are in place, they cost the city a mere $20,000 to $80,000 a year to run.

Every state in the union has had snow. Haleakalā volcano in Maui had a six-inch snowfall, and northern Florida got "clobbered" by a four-inch storm. The Texas Panhandle endured a storm that dropped twenty-five inches.[3] There are the other extremes. It's not unusual for Paradise Ranger Station in Mount Rainier National Park in Washington to log fifty-six feet in a single winter. Perhaps the record holder is Nagano Prefecture in the Japanese Alps, which can receive 125 feet in a single year.[4] Cities most likely to get drubbed by a good snowfall are municipalities like Buffalo, Rochester, and Cleveland near the Great Lakes, which are recipients of "lake-effect" snow.

Salt—124 pounds per person per year in North America—is the key to creating ice- and snow-free roads. Salt lowers the freezing point of water, which allows snow to melt at temperatures as low as 15 degrees F. Below that temperature, sodium chloride's effectiveness wanes. If Keding foresees frigid or icy conditions, his crews apply a prophylactic liquid "pretreatment" solution containing 35 percent salt, which helps prevent ice and snow from sticking to the roads. If it snows, he's a heroic, albeit unheralded, fortune-teller; if not, he's gambled away part of his budget.

To keep Minnesota's state and US roads and interstates free of ice, MnDOT uses almost a quarter million tons of salt each year. When I ask Stevens where such vast amounts come from, he reels off the names of a few states with underground salt mines. How big are they? Stevens recently ran a 5K in an underground Kansas salt mine.

A Marquette University study shows that salted, dry roads reduce crashes, injuries, and accident costs by 85 percent.[5] But the white stuff has its dark side. Drivers in the United States pay about $3 billion a year in rust repairs, and taxpayers another $10 billion in road and bridge repairs. If you want to become rich, invent an inexpensive alternative to salt—something that will melt snow with no harmful side effects. People have experimented with beet juice, pickle juice, and cheese brine—the downside being expense and an aroma similar to that of soy sauce, molasses, stale coffee, or all three.[6]

The Great Blizzard of 1888, which paralyzed the East Coast for five days, prompted changes in not just how snow was cleared, but how infrastructure was built. Fifty-five inches of snow pummeled some areas, and drifts up to fifty feet buried buildings, people, and animals. Fifteen thousand people were stranded on elevated trains in New York City. Four hundred people died across the East Coast. The storm prompted P. T. Barnum—who proceeded with two performances at Madison Square Garden—to comment that the storm might be an impressive show, but his was still "the greatest show on earth."[7] In the aftermath, many cities were divided into sections, and systematic snow-removal routes were developed. Plowing occurred during a storm, not just after. The storm set the wheels in motion for cities to bury telegraph, telephone, water, and gas lines and in ensuing years inspired the construction of below-ground subway systems.[8]

Sitting behind the wheel of a $200,000 plow truck gives me newfound admiration for the anonymous driver who plows my street. It also gives me anxiety. The controls mimic the complexity of a lunar landing module. On the mammoth right armrest are four joy sticks that control the height and angle of the front- and side-mounted plows. Touch screens on the dash control salt dispersion. Another screen shows GPS position. There are dozens of toggle switches for warning lights, headlights, and taillights. It's overwhelming and nerve-racking—and I'm sitting inside a warm, dry facilities garage.

Drivers need to be able to control all the above in a blizzard, driving thirty miles an hour, at 2 a.m., with who knows who or what on the road. Steve Kochendorfer, who drove a plow for thirty-one years for MnDOT tells me, "It becomes natural after a while."

During the 1991 Halloween blizzard, he got in a lot of practice. "It snowed so hard I couldn't get out of my own driveway," he says. "My foreman had to pick me up in a snowplow. I worked twelve-hour shifts for a whole week."

Kochendorfer talks about "gang plowing," when five or six trucks line up in staggered position with the precision of a Blue Angels squadron. "Get a good line going and you could clear the whole freeway in one good run. Then you'd turn around and get the other side."

It's a dangerous job. Kochendorfer tells about one truck driver who, forgetting to lower his truck box after emptying salt, hit an overpass. His truck box wound up in the middle of I-694. In 1992, the plow Daniel Jaramillo and his partner were driving was entombed by an avalanche. Jaramillo spent eighteen hours clawing his way by hand out of snow that one rescuer compared to concrete. His partner didn't survive. A plow operator in Idaho was killed when the snow thrower he was trying to unclog pulled him into the blades.[9] Utah plow driver Terry Jacobson survived a three-hundred-foot plunge down Spanish Fork Canyon after a semi-truck driver tried to pass him and hit his plow.[10] But more often than not when an accident involves a plow and a vehicle, it's the fifteen-ton plow that comes out on top. "You're not going to win if you hit a plow," Keding says.

Occasionally, plows are called into action to clear something other than snow. In 2015, the Iowa Department of Transportation used a snowplow to remove millions of mayflies from the Savanna-Sabula Bridge, which crosses the Mississippi. Plows have been employed to remove hail from the streets of Coon Rapids, Minnesota, and tumbleweeds from the roads of Bozeman, Montana. And the 1876 "Kentucky Meat Shower," which involved hundreds of meat chunks falling from the sky, proved that plows might need to be ready for anything. (Most people accept the "vulture theory," which postulates that the meat source was a kettle of vomiting buzzards flying overhead.)[11]

Knowing if and when to shag his crew out of bed is a gambler's life. Keding is on call twenty-four hours a day. In winter, he's constantly triangulating information from various meteorological sources. Weather forecasting is both a science and a black art; there's the Global Forecast System, as well as the North American Mesoscale, Nested Grid, Canadian, and European models. When I ask Keding which source is the most reliable, he replies, "None of them. It depends on which model the storm wants to abide by."

He explains tongue in cheek that squirrels can be better predictors of a harsh winter than meteorologists. "I keep an eye on them in the park outside my window. If they're running around, putting away food frantically in October, I know it's gonna be a bad winter. If they're just sort of lazing around, getting hit by cars, it's probably going to be a mild one."

Keding doesn't have to predict only the weather. Every April, he has to predict how much salt he's going to use during a winter seven months away. He needs to order X tons of salt; when winter approaches, he can buy anywhere from 80 to 120 percent of that amount, but regardless of the winter, he has to buy the minimum—and store it. In 2019, he ordered seven hundred tons.

The twenty-billion pounds of sodium chloride we spread on US roads every year negatively affects the health of our lakes, trees, aquatic life—and us. We can't see it, but every crystal of it goes somewhere. Heavier than freshwater, saltwater settles to the bottoms of lakes and ponds and impacts bottom-dwelling plants and aquatic life. It creates mutant butterflies and rainbow trout hatchlings. And it's a stubborn pollutant that does not easily go away. It attracts elk and deer to roadsides, which increases the incidences of human and animal fatalities. It accelerates bridge corrosion and eats away at vehicle brake lines and exhaust systems. "Everything is linked," Keding says. "Salt melts the snow, the salt impacts our land and streams, the altered environment affects us humans, humans determine how much salt to put on the roads. It's all tied together . . . even with something as simple as snow."

Our "dry road" policies, litigious mindset, and hurry-scurry pace all contribute to the problem. For decades, ice and snow were dealt with by plowing and spreading sand and cinders, primarily on curves, hills, and bridges. These abrasives often clogged storm sewers or filled ponds, but the problem was volumetric, not chemical. In 1960, when President Lyndon Johnson enacted a dry roads policy for the interstate system, state and local jurisdictions followed suit.[12]

As a result, salt use exploded and remains the preferred method for keeping roads drivable. While products like potassium formate are more environmentally friendly, they cost fifteen times more than sodium chloride. Part of reducing salt use involves reducing expectations. "Since people can drive sixty miles an hour down a road in June, they expect they can do the same thing in January," explains Keding. "If people would slow down, we could use a lot less salt.

"If you look at the Target parking lot down the road, you'll see it's stark white from all the salt on it—and the way the liability laws are written, companies have to do that. Again, people should change their expectations—and footwear. Don't wear three-inch heels when you go to Target. Wear boots. Less salt. Cleaner environment."

Is Shoveling Snow Good for You?
(Don't Hold Your Breath)

According to Stillwater's "Resolution Adopting Snow and Ice Control Policy," as part of my civic duty, I am responsible for "removing snow and ice from public sidewalks 24 hours after the snow and ice has ceased to be deposited." I get extra credit—and have less chance of my house burning down—if I clear out space around a fire hydrant. They don't care how I do it; they just want it done. A study by *Consumer Reports* tells me that if I were to purchase an $875 snow blower—instead of paying $40 a crack to "a plow guy"—my payback period in Minnesota would be a mere one year. The same period is three years in Boston, thirteen in Seattle, and "never" in Charlotte.[13]

Of course, there's another, cheaper option: a $29 aluminum shovel. Shoveling is grand exercise. It involves core, strength, and cardio training. It works the oblique, arm, leg, and back muscles. You can burn up to six hundred calories per hour—on a par with exercising on an elliptical trainer.

But it can also be dangerous, even deadly. One seventeen-year study, which analyzed 195,000 snow-shoveling-related emergency-room visits, showed 55 percent of the visits involved soft tissue injuries caused by "acute musculoskeletal exertion," 20 percent involved slips and falls, 15 percent involved being struck by a shovel, and 7 percent were cardiac related. Of the 1,647 deaths in the study, 100 percent—ALL of them—were related to cardiac arrest.[14]

The "whys" surprised me.

- People naturally tend to hold their breath when lifting and tossing a shovelful of snow. This "Valsalva effect" increases blood pressure and the chance of cardiac arrest.

- Cold weather causes small blood vessels to constrict, which also increases blood pressure.

- Shoveling involves strenuous arm exercise and places more stress on the heart than leg exercise does. People who excel on a treadmill don't always excel in shoveling snow.

- People tend to shovel in the morning before work, which often means they're in a hurry and their circadian rhythms are out of sync with the early morning workout.

If you're overweight, older than sixty, or sedentary, or if you smoke, have diabetes, or elevated cholesterol, your risk is even greater. The dangers are so great it prompted the cardiac team from one hospital in Billings, Montana, to visit hardware stores and place red heart stickers on shovels with warning signs about heart attacks.[15]

To minimize your chances of a "cardiac event," break the job into three or four outings and wait until later in the day when it's warmer and your circadian rhythms are more in sync with the task. You can also cover your mouth with a scarf to warm your breath, stay hydrated—and consider paying someone else $40 to spare yourself a $40,000 hospital bill.

PART V

SIGNS, LINES, AND LIGHTS

STOP!

Green Lights, Red Signs, *and* Roundabouts

I WAS MERRILY CRUISING DOWN THE INFORMATION HIGHWAY, guzzling facts and figures about traffic signals, when a YouTube video brought me to a screeching halt. The segment gave me the same sense of betrayal I'd felt upon learning there was no Santa Claus. The piece revealed that only 9 percent of New York City's three-thousand-plus pedestrian crosswalk buttons actually did anything; in Dallas, it's 0 percent. Those buttons people rely on to safely get to the other side are little more than sidewalk mannequins. All that sense of empowerment those little buttons once gave me—the ability to halt an entire city of traffic with one finger—dashed to the ground. I can only pray the stoplights they're attached to lead more upright lives.

To restore my faith in humanity, and traffic management, I turn to Steve Misgen, district traffic engineer for the Minnesota Department of Transportation (MnDOT). He sort of does. We meet after hours at a local microbrewery; Misgen has hair as white as the head on his beer and the solid demeanor of a good stout ale. He thoroughly knows his thoroughfares; he began working at MnDOT while in college and, thirty years later, is still waiting for the WALK sign to appear.

In regard to those little buttons, he explains that in dense urban areas, where traffic light timing is immensely complicated and traffic flows are fairly predictable, operable pedestrian crosswalk signals muddle the flow of vehicle traffic. To compensate, pedestrian traffic is always allotted adequate time as part of the overall timing cycle. In outlying areas, in small towns, in suburbs, and at many intersections in urban areas, those pedestrian buttons *do* indeed communicate with the traffic lights. But often, they're little more than thumb calisthenics devices for naive pedestrians.

After scrolling through the 294-page "Traffic Signal 101" manual Misgen loans me, I begin to grasp the complexity of traffic control. There's a *War and Peace*–length section on wiring, forty-six diagrams relating to left turns alone, a twelve-page treatise on metal cabinets. Traffic light timing is based on traffic counts, algorithms, warrants, and human sensibility. It's complicated.

Signal timing changes with the ebb, flow, and direction of traffic. Any given set of signals may run through three to ten different programmed cycles per day, and cycles can range from 90 to 250 seconds. Misgen presents a typical scenario. From 6 to 9 a.m., a set of lights may be timed to facilitate rush-hour traffic going east; from 3 to 6 p.m., they're timed for rush-hour traffic traveling west. "Those going against traffic are basically screwed—that's a technical term," he smirks. Midmorning, the signals may transition to an "off-peak" cycle, when the green-yellow-red cycle is shorter, and traffic flows more evenly in both directions. In between, there may be a special cycle to accommodate an increase in bus traffic from local schools or a lunch-hour rush from a nearby factory. Pedestrian volume, crash frequency, and proximity to railroad tracks also affect timing. When I ask Misgen what kind of person is attracted to his line of work, he responds, "Someone who likes working on a jigsaw puzzle the size of a city."

He explains how more than 90 percent of MnDOT stoplights can be remotely programmed from a central traffic-control center, where sensors, monitors, and cameras relay information about traffic flow, crashes, and backups. Four hundred cameras shuttle live images of roads and intersections to a bank of one hundred monitors.

Man, would I like to see that.

Misgen meets me at the door. I'm late because of a crash—a crash I later see live on a monitor in the traffic-control center. He gives me an eyeball tour of the immense room. The area has a Mission Control aura to it, but the task isn't guiding a single space module home safely. Instead, it's guiding the three-and-a-half million drivers and millions of vehicles that ply Minnesota roads. At the far end of the room is the 911 dispatch center for highway patrol; it's flanked by a cluster of people in charge of road maintenance dispatch who send out Highway Helper trucks to distressed vehicles and monitor the LED message signs that warn drivers of accidents or drive times. A radio broadcast booth includes a traffic DJ who gives live traffic updates every ten minutes during rush hour. To help get the word out about traffic jams and road conditions, MnDOT shares camera feeds with local television and radio stations.

I'm mostly interested in seeing the "signal operations group" in action. Some workers are watching computer diagrams depicting the multicolored ebb and flow of traffic throughout the state. Others are watching banks of live monitors, with images crisp enough to show facial expressions. It feels a little Big Brother–ish, but it's a

Steve Misgen shows the inner workings of a traffic-control box, similar to the one you see near every stoplight intersection—of which there are two million in the United States.

benign big brother. At one point, Misgen remotely rotates a camera and zooms in on the wheel of a Buick. The picture is so clear I can count the threads on a lug nut.

Misgen introduces me to Derrick Lehrke, a signal optimization engineer. Young and energetic, he's thrilled to be analyzing the two terabytes of information that flow through the computer network every day. He's especially excited about the new program that allows him to analyze the flow of traffic at more than 550 intersections.

Misgen's department gets five hundred to six hundred complaints a year, most of which are legitimate. They respond to them all. Lehrke tells the story of one driver who called complaining that he'd sat at a stoplight "forever." They located the video (they save four days' worth for each camera), located the vehicle, and confirmed that the man had, in fact, sat at the light forever . . . looking at his text messages and missing all the green lights.

The conversation swerves into the fast lane of *connected vehicle technology* (CVT). With CVT, vehicles use short-range radio signals to communicate to one another, to traffic signals, and to traffic-control managers. Misgen thinks that seatbelts, air bags, and improved car design have been monumental advances in helping people *survive* crashes, but he believes CVT will help *prevent* crashes. Every year, US roads are host to five million crashes and thirty thousand deaths; one study shows that CVT could reduce the number of crashes by 80 percent.[1]

How so? If a car is about to run a red light, drivers with CVT in the area would be alerted. If someone in oncoming traffic swerves into your lane to avoid an object, a warning would be broadcast to alert you. Sensors in the road could warn drivers of icy conditions. But there are likely other benefits. Traffic-control managers would be able to control traffic flows and cut drive times, fuel consumption, and emissions. In the event of a crash or medical emergency, "connected" vehicles containing health monitors could send information about injuries to medical personnel. The possibilities are staggering.

In the late 1800s, urban intersections had devolved into an anarchic tangle of horse-drawn carriages, bicycles, street cars, pedestrians, steam- and gas-powered vehicles, and frustration. Police were the primary form of traffic control, but this was labor intensive—and standing in a busy crossroad with only a whistle and white gloves for protection was not without its dangers.

In December 1868, traffic police near the houses of Parliament in London began using the first illuminated traffic signal.[2] Semaphore arms signaled drivers by day, and by night, a gas-powered light illuminated red and green lenses. The device leaked gas and within its first month of action exploded and injured the bobby operating it. The incident stalled the idea of newfangled traffic-control devices for decades.

Other concepts for stoplights evolved. In 1912, Utah police detective Lester Wire built an electronic, two-color traffic signal; two years later, the American Traffic Signal Company installed a version in Cleveland—but these, like most others, still required human operation.

At the turn of the twentieth century, twenty-five hundred cars a year were manufactured in the United States; by 1913, the output was nearly half a million.[3] That same year, four thousand people died in vehicle crashes, which meant thirty-four motorists died for every ten thousand cars on the road, compared with one and a half today.[4] Foreseeing the necessity of automatic traffic control at every intersection in urban areas, inventors hit overdrive. Initially, most signals were designed to stand in the middle of the intersection. Some were massive towers that included streetlights, street signs, police call boxes, even mailboxes. Others included an automatic traffic whistle that sounded "exactly like a policeman's whistle [that] makes both driver and pedestrian more alert."[5] Still others were low-profile mushroom-shaped affairs that could be driven over. Eventually, overhead lights suspended by wires, arms, or poles became the standard.

Heavily debated issues were the number of lights to include, whether or not they should flash, and the best arrangement. Some early signals showed yellow in both directions, enabling those stopped at red lights time to finagle shift sticks and teams of horses into gear. Different cities had different types of stoplights and rules. Eventually, national regulations were enacted dictating all lights be configured red, yellow, green from top to bottom, or left to right. Standards for lens size, turn arrows, and yellow lights were established. Red, traditionally associated with danger and historically flown by ships to indicate they were carrying explosives, was selected for "stop."

Pedestrian crosswalk signals were also developed. In one innovative twist, John Allen designed a pedestrian signal in 1947 that could be monetized with each click of a button; the words "stop" and "go" would be followed by the word "for" along with a product or brand name (STOP for Kleenex, GO for Coke).[6]

Perhaps no person knows more about traffic lights than retired fireman Willis Lamm, who resides in Stagecoach, Nevada, with his collection of seventy vintage stoplights—and sixty-five historic fire hydrants and forty vintage streetlights. Lamm loves the nostalgia, ingenuity, and design inherent in vintage traffic lights. He's particularly drawn to lights of the 1950s when stoplights, as well as cars, had "some degree of style and grace."

He's noticing a renewed interest in vintage traffic lights. "Oftentimes [historic] cities would spend vast sums on custom paving, expensive replica street lighting, antique benches, and decorative trash cans," Lamm explains. "But the stoplights installed could have been more suited to an industrial park."[7] Lamm tells me that some towns, like Winter Park, Florida, and Chapel Hill, North Carolina, are restoring existing lights or installing refurbished older ones to preserve their town's historic character.

Though the variables are mind-numbing, for most intersections to be considered for a traffic light, at least 600 cars per hour need to pass on the main line, 150 on the cross street. The volume of traffic at the intersection outside my front door doesn't come close. My corner gets two standard thirty-inch octagonal stop signs.

The world does not agree on much. An international standard of measurement? Nope. A universal language? Esperanto never caught on. But there are three universal truths when it comes to vehicles. First, virtually every car tire on the planet has a Schrader tire fill valve.[8] Second, the little arrow next to the gas pump logo near your fuel gauge points to the side of the car the gas cap is on. Third, in 1968 the Vienna Convention on Road Signs and Signals adopted two acceptable types of stop signs: a red octagon, and a circle containing an inverted triangle. Both signs contain the word "STOP" in the appropriate language.

The first stop sign installed in Detroit in 1915—a two-by-two-foot white square with black letters—looked nothing like the signs that stop us today. Its inventor, William Eno, was also credited with inventing the one-way street, the taxi stand, and—as we soon shall see—the traffic circle, most notably New York City's Columbus Circle in 1905. Known as the "Father of Traffic Safety," Eno never once drove a car, because he distrusted them.[9] By 1954, the US stop sign had evolved into one with white lettering on a red background to coincide with the red used in stoplights.

Yet despite the certitude of the sign's message—"STOP"— there are still seven hundred thousand accidents annually at US intersections.[10] These accidents account for 40 percent of all traffic fatalities. A high proportion of stop-sign accidents involve older drivers and those who complete the stop but are then hit by a vehicle "they didn't see." Drivers have difficulty obeying the four-letter word "STOP" under other circumstances too. Every year the Minnesota Department of Safety conducts a one-day survey asking bus drivers to report stop-arm violations; it averages six hundred breaches on that single day, translating into a hundred thousand annual stop-arm violations in one state alone.[11]

Despite their simplicity, stop signs come with hidden costs. Studies maintain that *each* stop sign–controlled intersection costs drivers between $110,000 and $210,000 a year in added fuel costs, brake wear, and travel time.[12] Yet what would be a better alternative? Perhaps we could look at options in a roundabout way.

I have a love-hate relationship with roundabouts. European countries do them with such grace—like two roads meeting, falling in love, and wrapping their arms around one another. US roundabouts seem to have been built using 3D printers and then dropped into place with a crane. But between 1990 and 2015, the number of roundabouts in the United States jumped from a mere handful to well over five thousand.[13]

Love them or not, they do reduce the number of accidents. They force drivers to slow down—typically to about fifteen miles per hour—and, since all vehicles enter the circle and move in the same direction, there are fewer head-on or T-bone collisions. No one's trying to "beat a light." Each year, more than nine hundred people—many of them pedestrians, bicyclists, or innocent drivers—die in crashes involving someone running a red light.[14] Roundabouts reduce collisions by 37 percent and fatalities by 90 percent.[15] A typical two-lane intersection, according to one study, includes twenty-four points of "potential conflict"; roundabouts have only four points—one at each of the merging points.[16] Roundabouts permit U-turns to occur with the natural flow of traffic, and since traffic flows in one direction, pedestrians can more safely cross. In many situations, roundabouts are quieter than controlled intersections.[17] Vehicles at roundabouts spend little or no time waiting and as a result consume less gas idling and "starting back up." Studies show roundabouts can reduce carbon monoxide emissions by up to 45 percent and fuel consumption by up to a third. Brake linings also like roundabouts. In already developed areas, they can be expensive to build, since their size often involves land acquisition. But in developing suburban and rural areas, roundabouts can be less expensive to install than traffic lights and require little maintenance beyond plowing and sweeping, making them increasingly popular in these areas.

The costs of traffic control, road construction, and road maintenance are staggering. MnDOT employs five thousand people, has a budget of $3 billion, and oversees 144,000 "lane miles" of road. Installing a set of traffic lights can cost from $150,000 to $500,000. Factoring in land acquisition, the cost of a roundabout can push $2 million. But the return on investment per year is huge. One of Misgen's directives is to create better traffic control to minimize the number and scale of road construction projects. He told me his department eliminated the need for one $10 million interchange project simply by retiming the traffic lights. "It's estimated that for every dollar MnDOT spends on traffic control," Misgen touts, "there's a savings of anywhere from $40 to $200 to the public. Fuel is saved, people waste less time, there's less pollution." Everyone wins.

WALK THE WALK

Seven Things You Don't Know About Traffic Lights (and Maybe Shouldn't)

1. To determine how long a stoplight will remain yellow, divide the speed limit by ten; in a fifty-five-mile-per-hour zone, the signal will stay yellow for about five-and-a-half seconds. Though the actual formula looks like something from Einstein's chalkboard, this calculation will get you within half a second. Regardless of speed limit, most yellows last at least three seconds.

2. All traffic lights have the added precaution of a momentary "all-red" phase in which lights are red in both directions. The average length ranges from one-and-a-half to three seconds.

3. Most traffic lights complete a cycle in one-and-a-half to two-and-a-half minutes, though some can take more than four minutes. The traffic light where you "waited forever" probably took no more than three or four minutes to cycle.

4. The inductive loops embedded in the pavement that signal to traffic lights that you're approaching are triggered by the metal components of your car. An all-plastic car or an elephant won't trigger the light. (Oftentimes, bicycles won't trigger the mechanism either.)

5. Crosswalks are timed to provide four to seven seconds for pedestrians to begin crossing plus one second for every three-and-a-half feet of street they need to traverse— a countdown usually indicated by a flashing hand.

6. Some pedestrian crossing buttons can be programmed to provide extra time when pushed for a certain length of time or in a certain pattern; this is primarily done in areas with a high population of senior citizens, kids, or people with physical disabilities.

7. Emergency vehicle preemption (EVP) systems will trigger a green light when the strobe lights of approaching emergency vehicles are detected. It will "freeze" the green light until the vehicle passes. This can be irritating— unless you're the one having a heart attack—but it reduces intersection crashes by 70 percent and increases response times by 25 percent.

In 2017, 939 people were killed by vehicles running red lights—the highest toll in a decade. To reduce your chances of being T-boned, you can take two defensive measures that will protect you from distracted drivers. First, if you're the first car in line, wait for a second or two and look both ways after your light changes before proceeding. And second, be prepared to stop when approaching or entering an intersection, even if you have the right-of-way. This has saved my life twice.

ROAD LINES AND SIGNS

The Language of
Signs *and* Dashes

DRIVE TWO MILES AND, ACCORDING TO A RECENT STUDY, YOU'LL make four hundred observations, forty decisions, and one mistake.[1] And then come the next two miles. And the next. Many of your observations, decisions, and mistakes have to do with lines and signs—two everyday things that, like actors in a Rudolph Valentino movie, are silent yet clearly tell their stories.

If you're driving a stretch of Route 99 in California and encounter a marker reading "The Doctor June McCarroll Memorial Freeway," you may presume she saved someone's life along this stretch of desert roadway or perhaps had volunteered to pluck up litter along it.

No. In 1917 Dr. McCarroll and her Model T were unceremoniously run off the road by a distracted driver. "It did not take me long to choose between a sandy berth to the right and a ten-ton truck to the left!" she explained.[2] Determined to prevent this from happening again, she began campaigning for painted white lines down the centers of local roads. When officials scoffed, Dr. Mc-

Carroll took matters into her own hands—literally; she grabbed a paint brush and bucket of white paint (by some accounts, cake flour paste) and painted the lines herself.[3]

For the next seven years, she relentlessly campaigned for center lines, and in 1924, the state of California instituted the practice. Others soon followed. Nothing is without controversy, however. Edward Hines, chairman of the Wayne County Board of Roads near Detroit, claimed he came up with the idea while following a white line left by a leaky milk wagon.

"I tell people, 'It ain't just paint,'" says Ken Johnson, state of Minnesota work zone, pavement marking, and traffic devices engineer. "Pavement markers are a language that communicates to the driver what they need to do and where they need to go."

Johnson explains that all states follow the guidelines laid out in the federal Manual on Uniform Traffic Control Devices, lovingly nicknamed the MUT. When Johnson pulls out his eight-inch-thick edition and drops it on the table, the stapler and paper clips quake. "We want someone from Florida to be able to drive in Alaska, Hawaii, and Virginia with no surprises."

He and his co-worker Ethan Peterson explain that colors and patterns speak to us. Yellow lines separate lanes of oncoming traffic; white lines separate same-direction traffic. Perpendicular lines make us aware of crosswalks, yield zones, and stop zones. Most solid lines tell us to stay in our lanes, long dashes to pass with care. Short dashes tell us our lane is about to merge with another or head onto an exit. An *X* tells us railroad tracks await ahead, diamonds that the lane is for high-occupancy vehicles, and chevrons that bikes and cars are expected to share the road (*sharrows*). Blue is reserved for accessible spaces and green for bike lanes.

This is a sign language most of us understand.

Ethan Peterson takes a close look at samples of the paints, tapes, and reflective beads his department uses to make the road lines that talk to us every day.

But what gives road paint the cojones to stand up to forty-ton trucks, hail, road temperatures that swing from 50 below to 150 above, torturous snowplow blades, torrential downpours, Camaros burning rubber, and sand that gets ground in by the minute?

Lines and road markings are made from a variety of paints and tapes, all of which are determined on the basis of budget, surface, projected longevity, temperature, humidity, average daily traffic, and how long a road can be taken out of commission during "striping." City, county, state, federal, and private roads come with different budgets and regulations. A small town with a small budget probably goes with good old latex paint, which costs twenty cents a foot and lasts one to two years. A rich city with a fat budget—and brand new concrete streets—will likely go with newfangled thermoplastic or polymer tapes, which cost $3 to $6 a foot.

But again, it ain't just paint—or tape. For pavement marking to do its job at night and in the rain, it needs to have *retroreflectivity*, or something in or on the paint or tape to reflect light from the headlights back to the driver's eyes. The most common way to boost reflectivity is with beads of glass or ceramic, often made of recycled bottles. The sand-size beads reflect best when they're embedded with about 40 percent of their diameter above the surface. Too deep, and they won't reflect; too shallow, and they pop out. One Dutch company invented a "beadless" photo-luminescent paint that absorbs sunlight by day and radiates light by night.

By now—if you haven't raced out the door to check the retroreflectivity of the lines on your street—you should be asking, "Don't snowplows and other abrasive things beat those tapes and little beads to smithereens?" Yes. Yes, they do.

Peterson explains this is why when many roads are built or rebuilt—especially in snow-prone states—shallow recesses are cut for those paints and tapes to nestle into. This is particularly important for tapes that can be ten times thicker and cost ten times as much as paint. The good news is they can last up to ten times as long, which is considerably longer than the first reflective line tape developed in 1939 by 3M, dubbed "3M's friendly tape" because it would often strip free and wave at passing motorists.[4] The life expectancy of road paint also doubles when recessed. Grinding grooves isn't cheap, but in the long run it saves money, work, and lives.

Reflective beads come in a variety of sizes, materials, and qualities; some are good at reflecting light through water, others are better in dry conditions. But both types are included in most surface markings. The reflective beads are blasted into the paint, and the paint acts as the "glue" to hold the beads in place.

Tapes have reflective beads or embossed patterns already in place. Installing the preformed polymer tapes involves removing the backing and then pressing them onto the surface or into the recess. Thermoplastic tapes are installed by positioning them, then heating them with a torch. Arrows, symbols, and words are often made of thermoplastic tape, but many continue to be painted using enormous stencils.

Pavement markings are worth every penny of the $2 billion a year US departments of transportation spend; one recent study shows that just adding edge lines and center lines to rural two-lane highways reduces crashes by 36 percent.[5] Thank you, Dr. McCarroll.

I drive a Honda truck with a Lane Keeping Assist System (LKAS) and Road Departure Mitigation System (RDM). If I drift out of my lane, the LKAS vibrates the steering wheel like a jazz drummer, and the RDM tries its best to keep my truck between the lines while flashing "STEERING REQUIRED."

I ask Johnson and Peterson what role road markings play in guiding smart cars. "The radar, cameras, and sensors are looking at a lot of different things," Johnson explains. "Surely the lines play a large role but so do signs, curbs, trees, guardrails, and moving objects." Right now, lots of conversations are taking place between auto manufacturers and government bodies. "There's a lot to work out," Peterson says.

There's another thing that makes my steering wheel shake, rattle, and roll: rumble strips. We've already seen that line paint can reduce crashes by more than a third, but "audio tactile profiled markings" (as they're called in the DOT biz) top that. Rumble strips reduce head-on collisions on rural two-lane roads by up to 50 percent and on two-lane urban roads by up to 91 percent. Shoulder rumble strips reduce run-off-road injury crashes by about 25 percent.[6] They cost between $3,000 and $3,500 per mile to install and are made by grinding a series of half-inch-deep, five-inch-wide grooves at regular intervals. "They're one of the lowest cost treatments you can do for improving safety," Peterson says.

Not everyone loves them, particularly people who live within hearing range—which can be up to two miles. The same *BRRRT, BRRRT, BRRRT* that wakes up drivers, also wakes up homeowners. People complain of sleep deprivation and ruined lives. One Wisconsin resident complained that the noise was so nerve-grating he had to take his wife to the emergency room to treat her high blood pressure.[7]

In response to complaints, "mumble strips" have been created. They're still ground into the pavement, but in more of an undulating pattern that creates a sound less irritating to neighbors, yet still irritating enough to wake up inattentive drivers. Peterson compares the noise to driving on gravel.

Whether you call them rumble, mumble, or grumble strips, they save lives. Johnson points to letters taped to his door. One is from a woman who was so blinded by wind and rain (and afraid to pull over for fear of being rear-ended) that she used the sound of the rumble strips to guide her to the nearest exit. Another is from a sleep-deprived, boat-towing fisherman who fell asleep at the wheel, with his son in the back, but was jolted awake by rumble strips. "I know the department of transportation gets a lot of crap for road construction and things," the letter explains, "but I owe them big time for this one."

Lines are the first device for keeping you on the road, rumble strips the second, and guardrails the third. "Most people consider guardrails hazards," says Johnson. "But they're less hazardous than the things you'd run into or over if they weren't there."

Lines, bumps, and rails tell only half the story; let's look at their literary counterparts.

How hard is it to put up a road sign? You dig a hole, stick a post in the ground, and the world becomes a safer place, right?

Well . . .

First you need to determine whether a sign is really, really, really needed because if you put up too many superfluous signs, drivers start ignoring ALL signs. And, as we'll soon see, many signs already have no effect on driver behavior—whatsoever.

You need to consider the amount of vehicle, bike, and pedestrian traffic, the speed they're traveling, sightlines, and previous accident rates.

You need to determine sign color and geometry because people recognize a sign first by its shape, second by its color, and third by what it says. Color and shape are usually dictated by federal standards, but when you get to size—whew!—break out the computer and manuals. Because, on freeways for example, for a sign to be legible, the letters need to be one inch high for every thirty feet of desired legibility, meaning most letters are sixteen inches tall.

You need to make sure the sign has adequate retroreflectivity so it can be seen at night—critical since the fatality rate is three times higher at night than during the day. You need to know that human eyesight peaks at the age of twenty, and after that it's all downhill. Visual acuity is halved every twelve years—so at age thirty-two your eyes need twice as much light to read the sign you read in your freshman year of college, and at age forty-four, four times as much light.[8] And it's no wonder eighty-year-olds should strongly consider hanging up the keys at night.

Then you need to figure out *where* to install the sign. For informational signs, you need to give drivers enough time to read the sign, understand the sign, and make a decision before reaching the next sign. On a high-speed road, that often means spacing signs eight hundred feet apart. A hierarchy determines which signs get put where.

Then you need to figure out how to mount it: post, overhead arm, or overpass. If it's on a post, you need to spec what kind of breakaway supports to use so if a vehicle hits it, the post or posts will shear off, inflicting minimal damage to vehicle, driver, and post. Yet you need to be sure the sign can withstand winds of up to ninety miles per hour.

Then you can dig a hole and install the sign. But it will cost you; by the time you've had the sign made, shipped, hauled to the site, and mounted using the right posts and hardware, even the simplest sign will cost a grand. And don't be so cocksure the world is going to be a safer place.

I'm sitting with Josie Tayse, state signing engineer, and signing supervisor Rick Sunstrom during another visit to the Minnesota Department of Transportation. Tayse is a generation younger, four inches taller, and has better cheekbones than Sunstrom, as well as a third as many years on the job as his fifty-one-and-a-half. But they share an absolute passion for signs.

Sunstrom, who resembles Harry Caray and loves baseball roughly as much as the legendary Cubs announcer, remembers making signs the old-fashioned way: "We'd cut the letters and numbers out with scissors and figure out the design and spacing, and I loved it because I love working with numbers."

Tayse grew up on a farm and always enjoyed fixing things and figuring out how they worked. She loves the variety of her work. "One day I'm talking with the state legislature, the next with some citizen group about a new school crossing sign, the next with maintenance guys, or another engineer, or with a business," she explains. "I get a little bit of everything."

The duo discusses some of the basics of the sign world. That there are three type of signs: *regulatory signs*—which call out traffic laws and restrictions (STOP, YIELD, No U-Turn) and are the only ones you can get ticketed for; *warning signs*—which alert drivers of the unexpected (curves, merges, deer, ice); and *guide signs*—which

Rick Sunstrom—with fifty-one-and-a-half years of sign experience under his belt—explains how objects in your windshield are larger than they appear. The freeway speed sign, like that in the foreground, is one-and-a-half times larger than your front door.

show destinations, distances, directions, services, and points of interest.

They discuss the nuances—such as the facts that most signs have a twelve-year life span, and that some breakaway signs are designed to fly over a car, others under. They talk about how sign reflectivity has improved so much over the past thirty years that they've removed almost all sign lighting.

Archaeologists have unearthed massive granite "milestones" placed by the early Romans to help travelers navigate the 250,000 miles of roads crisscrossing the empire; each displayed the distance between it and the "Golden Milestone" at the center of Rome.[9] In that sense, all roads *did* lead to Rome.

In the 1600s, British law mandated that all parishes erect signs indicating which roads led to which cities. The first bicycles—fast, quiet, hard to control—prompted the installation of many signs; some warned bicyclists of dangers ahead, while others warned pedestrians of dangerous bicyclists ahead. Those warning of steep hills bore a skull and crossbones.

The evolution of the horseless carriage spurred modern sign making into high gear. Automobile clubs began posting directional, mileage, and warning signs. But there was no uniformity (and surely no retroreflectivity). In 1923, state officials got together to come up with a plan to standardize signs. According to urban legend, they proposed that the number of sides a road sign would have would be commensurate with the hazard at hand.

Circles (having an infinite number of sides) would be used for the most lethal hazard of the day: railroad crossings.

Octagons would be for stop signs.

Diamond-shaped signs would call out warnings.

Squares and rectangles would be for regulations and speed.

Equilateral triangles would indicate the need to yield.

(The *pentagon*, added later, focused on school safety and crossings.)

Eventually colors were added to reinforce the message. *Red* for stop, *yellow* for general warnings, *black and white* for regulations and ordinances, *blue* for services like hospitals, hotels, and restaurants, *orange* for construction zones, *brown* for recreational areas, and *fluorescent yellow* for pedestrian, bicycle, and school bus warnings.

The first Manual on Uniform Traffic Control Devices (the MUT, which we've already met) was developed in 1935 to standardize signage; uniformity now reigns across all fifty states.

Retroreflectivity for nighttime sign recognition is critical. Signs are tested regularly using either a $10,000 retroreflectometer or a human inspector "at least 60 years old with 20/40 normal or corrected vision," driving an SUV or pickup truck that's model year "2000 or newer."[10]

HACKS & FACTS

Dashed Illusions

Close your eyes and guesstimate the length of the lines and spaces on a typical highway. Wrong. The dashes are almost always ten feet long and the spaces between them are forty. Everything seems more compact at high speed.

Faded or blocked signs are the number one complaint of drivers and the third leading cause of crashes.[11] Yet the hidden underbelly of signs is that they often don't change driver behavior or minimize crashes—the two ultimate goals of signs. As one manual explains, "It appears that most signs fall into a category of hope—hope they do some good and an expectation that at least they don't do any harm."[12]

Studies show that people do pay attention to the bright yellow warning signs that alert them to real and constant dangers. The curvy arrows that warn of upcoming curvy roads reduce road departure crashes by 25 percent and, if the road also has chevron signs on the curve, by 50 percent. But those big yellow signs that warn of infrequent occurrences or dangers—DEER CROSSING (where 99.9 percent of the time deer aren't crossing), FALLING ROCK (where 99.999 percent of the time rock isn't falling), and even PEDESTRIAN CROSSING (where pedestrians actually *do* cross a lot)—don't slow drivers down a whit. They're so ineffective that many departments of transportation are removing static warning signs. Or adding flashing lights that are activated only when an actual danger is present (motion detectors that detect deer, road sensors that detect ice, push buttons that signal a real person has pushed the button).

Speed limit signs present another conundrum. Speed limits are established by 85th percentile speed—"The speed at or below which 85 percent of all vehicles are observed to travel under free-flowing conditions."[13] Studies show that drivers don't pay attention to speed limit signs but rather to subjective things like road conditions, road width, curves, and the presence of parked cars and pedestrians. Drivers also establish their speed based on "the feel" of the road. Vehicles traveling at a uniform "natural" speed lead to fewer collisions and fewer drivers darting through traffic. Speed limit signs designed to unrealistically slow down traffic seldom work, and again, signs that are repeatedly ignored lead to driver "sign apathy." Test after test shows that if the majority of drivers are driving fifty miles per hour on a road marked forty-five, they'll still drive fifty after that sign is changed to thirty-five. "Wishful" speed limit signs make us all perpetual lawbreakers.

Tayse and Sunstrom not only have to deal with sign installation, they have to deal with sign pilfering—ironic since signs in some states are made in prisons. In the past, the most frequently stolen signs were those listing city name and population. "We couldn't keep the 'MINNEAPOLIS, POPULATION 432,000' sign in place," Sunstrom says. "For some reason, it was a fad. College kids would get drunk, unscrew them, and put them in their dorm rooms." Anything with "69" on it—with its sexual innuendos—gets stolen. Four Highway 69s in the United States have been renumbered because of perpetual sign theft. Signs for mile marker 420 (the symbol for April 20—the international pot-smoking day) were stolen so frequently in Colorado that officials replaced them with 419.99 mile marker signs. The legendary nature of these signs has now made them a target of theft as well.

$750,000 Speeding Tickets

About forty-one million speeding tickets are issued in the United States every year, which means your chance of receiving one is about one in six. Fines range from $10 for minor infractions in North Dakota to $2,500 for driving faster than eighty miles per hour in Virginia.[14] The average ticket amounts to $150—which may seem like a lot—until we visit Finland or Switzerland, where speeding fines are based on the offender's income. In one case, a Swedish driver clocked at 180 miles per hour driving through Switzerland was fined $750,000.[15]

A fine like that may help you ease off the pedal in the future. So might a few other facts:

- Speeding is a contributing factor in 26 percent of all traffic fatalities.

- The first day and the last four days of the month are the five days during which the most speeding tickets are issued.

- Only 3 percent of all speeding tickets are issued to drivers going ten miles per hour or less over the speed limit. BUT, the highest percentage of speeding tickets, 25 percent, go to those driving twelve miles per hour over the limit— a fine line.

- The hour or two following morning rush hour and the traffic lull between 1 and 3 p.m. are the times you're most likely to get a ticket.[16] Law-enforcement officers hesitate issuing tickets during rush hours since traffic stops can create gawker backups and accidents.

- A whopping 33 percent of Lexus ES 300 drivers have received a speeding ticket, while only 3.2 percent of Buick

Encore drivers have. Other "speedy cars" include the Nissan 350Z (32 percent), Dodge Charger (32 percent), and Volkswagen Jetta (31 percent).[17]

- Men are 50 percent more likely to be issued a speeding ticket than women.

25

STREET NAMES AND NUMBERS

Stravenues *and* Psycho Paths

OLE WAKES UP ONE MORNING, LOOKS OVER AT LENA, AND REAL-izes something is terribly wrong. He dials 911.

"Operator, you gotta help me. My wife is unconscious. I need an ambulance right away."

The dispatcher asks, "Where do you live, Ole?"

"On the corner of Rhododendron and Eucalyptus," he responds.

"Could you please spell those for me?" the dispatcher asks.

After a long silence, Ole replies, "How about I just drag her over to First and Oak and you pick her up there?"

Thank God the dispatcher didn't use the word "odonym"—the term for the name of a street.

Street names and numbers are important. Without them we wouldn't get our mail, pizzas, or ambulances. But where do they come from and why?

Once upon a time, road names were actually descriptive. *Avenues* were the grandest of roads, while *streets* ran perpendicular to them. *Boulevards* and *parkways* referred to wide, tree-lined roads.

Lanes were usually narrow, *drives* were winding, and *terraces* followed a hill or slope. *Places* and *courts* usually led to dead ends. There were even *stravenues*, the name given to the diagonal roads running through Tucson.

"Names do not necessarily have significance, aptness, originality, aesthetics, history, or any other aspect resulting in a meaningful street or park name," laments Donald Empson, author of *The Street Where You Live: A Guide to the Place Names of St. Paul.* "We have such misnomers as College Avenue with no college, Palace Avenue with no palace, Ocean Street with no ocean, and Hunting Valley Road with no valley or hunters. The selection of place names has often been, almost without exception, careless."[1]

And given the top ten street names in the United States, I suspect Empson is correct about originality:

No. 1: Second
No. 2: Third
No. 3: First
No. 4: Fourth
No. 5: Park
No. 6: Fifth
No. 7: Main
No. 8: Sixth
No. 9: Oak
No. 10: Seventh

Tree names start coming in hot and fast after No. 10. We don't hit the first human—Washington—until No. 17.

About half of all US cities have streets with a numerical sequence—most commonly in the downtown core.[2] About 80 percent of those cities use "number" streets in only one direction to avoid confusion. Vermonters—in line with their independent thinking—have only 6 percent of their cities with numbered sequences. Utahans—apparently more numerate—use number sequences in 92 percent of their cities.

Themes are common. Washington, D.C., has one street named after each of the fifty states. In Almere, the Netherlands, streets are named after dances (Tangostraat, Salsastraat, Rumbastraat), movie stars (Natalie Woodpad, Grace Kellystraat, Peter Sellershof), musical instruments, and fruit. One area of Grantham, England, has roads named after golf courses.

Empson much prefers street names that are descriptive, have local flare, or, at the very least, have humor. "Where was the developer with the creativity to name his addition the Urinary Tract, with streets such as Kidneystone Lane and Prostate Drive?" he asks.[3]

When I meander down our driveway, I find myself at the corner of Olive and Fifth. "Fifth" doesn't conjure up much excitement, and "Olive"—regularly usurped by the more succinct "Oak"—has a bit of an inferiority complex.

Before finding a street with even a remotely interesting name, I need to hop in my car and pass streets named after trees, dead city founders, developers' daughters, and presidents. If I head cross country, I start encountering the good stuff. I pass Total Wreck Lane, Farfrompoopen Road, Wayne's World Drive, and Horneytown Road (marked by a sign stolen so often authorities have installed a GPS tracking device). I encounter punny names like Psycho Path, Divorce Court, and Lois Lane. I drive down the painful Weiner Cutoff and lackadaisical Goodenough Street.

I see street signs that spout philosophy such as Peace and Quiet Road and When Pigs Fly Drive. I drive through intersections where Back and Fourth Avenues and Ho and Hum Roads cross.

While most new street names are suggested by developers, they do need to go through a review process. Police and fire departments want to make sure the names are unique enough, intelligible enough, and aligned with the "theme" of the area so emergency responders know immediately where to head. The post office and public works department make sure there are no duplicate names or double entendres. Catherine Nicholas, a San Diego developer, explains, "Many developers try, often successfully, to name streets for themselves, their partners, wives, mistresses, and children." Caution should be observed since the name of a street can affect the sale and even value of a property. "The street name can be a real turn-off or advantage," Nicholas says.[4] And they can be controversial. People for the Ethical Treatment of Animals (PETA) recently asked the city of Caldwell, Idaho, to change the name of Chicken Dinner Road to one that "celebrates chickens as individuals, not as beings to kill, chop up, and label as 'dinner.'"[5]

The author pauses at a street sign in Tulum, Mexico, realizing he may have lost his way.

Changing a street name is enormously complex, since it requires residents and businesses to change stationery, business cards, advertising, and contact information. Nicholas says it's "a very, very big deal" and "very, very rarely approved." Even in Atlanta, where people joke that half the streets are named "Peachtree" and the other half are declarative sentences, name changes are rare.

*As early as 1512, buildings in the area of Paris around Notre-*Dame Cathedral were given numbers, more to designate who owned what, rather than who lived where. House numbering schemes quickly became popular; they simplified taxation, law enforcement, and inscription.

But the address system in early American cities was chaotic. Even when a block was given a designated sequence of numbers, as buildings were torn down and replaced, some were numbered according to the order in which they'd been built, and others were given fractional addresses. Some property owners considered an address "private property" and felt they were free to do with it what they wished. The accurate naming of streets and numbering of houses became of paramount importance in the 1860s with the introduction of Free City Delivery mail. Two basic numbering plans emerged.

The "decimal" or "Philadelphia" system assigned one hundred numbers to each block, with odd numbers on one side of the street (usually south and east sides) and even numbers on the other side (usually north and west sides). House numbers were usually dished out by twos or fours and then "reset" to "00" at the start of the next cross street.

In a "continuous system" the numbers just kept going up in sequential order regardless of cross streets. Sometimes these numbers were determined by distance from an axis street, and other times by a complicated formula. In the parts of Minneapolis that use the decimal system, I know that when I'm at 2204 Zenith, I have to walk only a block to get to 2304 Zenith. But in St. Paul, which mostly uses the continuous system, I may need to trudge two or four blocks. Many cities are a hardscrabble mix of the two systems.

The inimitable Carmel, California, has its own unique address system—there is none. Houses are identified by name ("the Harrison Hamlet"), assessor's parcel number, or description ("east side of Lincoln, four houses north of 8th"). Most residents pick up their mail at the post office. It's a social thing. In 1926, the city council passed a resolution that forced residences to have house numbers; soon after, they were voted out of office and the measure was repealed.[6]

If Empson could step back and take a clean swing at it, he'd "let the names be creative and different from those in other cities. Let there be an official policy that the merest suggestion of any place name containing the words *oak, hill, fair, pine, mount, wood, ridge, dale, crest, grove, hurst, land, park, edge, glen, high,* or *valley* will automatically incur a fine of not less than $100 and at least a month in the workhouse."[7]

HACKS & FACTS ## Interstate Logic

The Dwight D. Eisenhower National System of Interstate and Defense Highways, or the Interstate Highway System, consists of more than forty-eight thousand miles of freeway. Built solidly enough—as the name infers—to bear the weight of heavy military equipment crossing the country, each mile contains more than three million tons of concrete. All but five state capitals are linked to the interstate system (sorry, Juneau, Alaska; Dover, Delaware; Jefferson City, Missouri; Carson City, Nevada; and Pierre, South Dakota).

There's a logic to the numbering system.

- Even-numbered interstates run east and west, with the lowest number starting in the south.

- Odd-numbered interstates run north and south, with the lowest number starting in the west.

- Sections of interstate that loop around cities are given three numbers with an even starting prefix (I-494). Spur routes—which don't complete a loop—are given three numbers with an odd-numbered prefix (I-587, I-110).

- Exit numbers are keyed to the number of miles they are from the western or southern border of a state.

Contrary to urban legend, interstates weren't designed with a one-mile stretch of straight roadway every five miles for military planes to take off from and land on.

GRAFFITI

Making a Mark
on the World

I'VE MADE ARRANGEMENTS TO MEET WITH GRAFFITI ARTIST Irvin at a small corner café in the Belleville area at the outer rind of Paris. I'm not sure what he looks like, but when I spot a laid-back, twenty-something dude with a hand-rolled cigarette and wearing a T-shirt with Mickey Mouse holding a can of Adidas spray paint, I feel I've narrowed the field considerably.

I come equipped with a list of questions, pen, and audio recorder; he comes with a backpack that rattles as if it holds ten cans of spray paint.

"Let's paint," he grins, nodding his head toward the alley behind the café. "We'll talk while we paint." It's broad daylight, my French is *très petite*, the fine for being caught is 3,000 euros—and my finest spray paint work to date consists of a small, black wrought-iron table. Yet . . .

As we walk past the graffiti-covered walls of the alley, I ask him whether he's sure this is okay. He smiles, opens his backpack containing ten cans of spray paint, and says, "The first rule of graffiti is to work BIG. Small is no good. BIG. Here, shake this can."

The author conducts hands-on research—and makes a name for himself—
while exploring the wide world of graffiti in Paris.

Little did Illinois paint salesman Ed Seymour realize in his
quest to create a better way to paint radiators that his newly invented spray paint would soon become the medium of choice for
graffiti artists around the world.[1] Concocted in 1949, then per-
fected in 1953 with the invention of the clog-free nozzle, spray
paint was quickly grasped by artists as unique. One could create
bright solid colors in seconds without drips and thick texture. It
could be used with stencils and masks. A sweeping hand gesture
left a sweeping path of paint. Unlike brushes and rollers, spray
paint could create precise images on rough surfaces—like brick
walls, freight trains, or roll-down security doors. And there *was*
just something that beckoned the works to be BIG.

When most of us think of spray paint, we think of $2.99 Kry-
lon at Home Depot. Serious graffiti artists use Montana and Iron-
lak, which can cost $10 and up per can. Some aerosol cans, issued
in limited editions, are considered works of art themselves and can
sell for upward of $300.

The "brushes"? Interchangeable tips with different spray patterns. *Needle caps* dispense quarter-inch-wide pencil-like lines. *Fat caps* spray eight-inch-wide swaths. *Ultra-line* caps can shoot a trajectory of twelve feet for hard-to-reach places. *Calligraphy caps* create fuzzy-edged lines.

For an artistic style that employs vibrant blues and yellows, there's a lot of gray when it comes to defining graffiti. Where is the line between street art and graffiti? Where does vandalism end and beautification begin? Does it need to make a statement or involve an element of risk? If a young Picasso had painted *Guernica* on a back wall of the Museo Reina Sofía one night in 1937, would it still be viewed as the priceless antiwar painting it is today? Or would it have been whitewashed over the next day?

Because it's hard to define graffiti, it's hard to discuss its history. "Graffito" and its plural form "graffiti" come from the Greek word *graphein*, which means "to write." Originally, graffiti referred to the inscriptions and figure drawings found on the walls of ancient ruins such as the catacombs of Rome. Archaeologists can point to an image, still visible at Ephesus, containing a heart, footprint, number, and woman's profile, as an early example of graffiti. Its purpose? To show the way to a brothel.

Throughout the ages, graffiti has used both images and words to make political or social statements. It's difficult to escape the poignancy of the words scrawled on a wall at Verdun:

AUSTIN WHITE, CHICAGO, ILL., 1918 AND 1945.
THIS IS THE LAST TIME I WANT
TO WRITE MY NAME HERE.[2]

Perhaps the most iconic graffiti image of all time—Kilroy, with eyes, nose, and fingers peering over a wall—was emblazoned onto the world's buildings and conscience during World War II, serving as a symbol for progress made by the Allies.

Until the 1960s, graffiti wasn't necessarily considered an art form. The emergence of "modern graffiti" can't be attributed to any one person, place, or thing. Vanguard artists have always been eager to experiment with new materials—and spray paint opened a new world of possibilities. The rise of urban hip-hop, British rock and roll, gang culture, and general rebellion were other elements.

Irvin's and my first step is to create an imperfect blank slate. We use white spray paint to cover the dark lines of the word "Rainer" and the cheek of a five-foot Moroccan-style face. I ask Irvin about graffiti etiquette. When is it okay to cover up another person's graffiti? He explains that the work of recognized artists receives an undefined tenure. A hastily scrawled name over an established, respected painting is considered *dissing*, a sign of disrespect. When asked whether there are *any* off-limit surfaces, he explains that historic monuments, churches, mosques, and other religious structures are usually respected. Few graffiti artists go after single-family residences. Commissioned street art is usually left alone. And only the most disrespectful graffiti artists will *tag*, or add their signature to, a statue or other object that is already a work of art. Ikaroz of Stockholm sees graffiti as "anarchy with rules."[3]

In the specialized world of train tagging, the unspoken rule is to leave the tracking number of the car exposed. One railyard employee explained the uneasy truce: leave the numbers exposed so we can track our cars and we might turn our heads a little; start covering them up and the hammer comes down.

As with most movements, a shared language has evolved. *Wild style* refers to an artist's name with letters so stylized it takes serious cerebral contortions to read. *Throw-up* and *bubble art* refer to the large curvaceous lettering I'm using in my Parisian graffito; *bombing* to a blitzkrieg attack on a surface. *Angels* are dead graffiti artists. *Toys*—an acronym for "trouble on your system"—are incompetent or, at best, inexperienced artists. When I ask Irvin if I would be considered a toy, he nods with conviction.

Irvin grabs paper and pen and asks me what I want to paint. Pictures? Words? Both? "Spike," I suggest. He sketches for a few minutes, fiddling with style, proportion, lowercase versus uppercase, then picks up a can of COLORZ black paint. In large sweeping strokes, he establishes the outline of the *S*, mortising the bottom around an existing letter. Then *P*. Then *i*. Meanwhile, I look up and down the alley in search of pissed-off shop owners or police. "Keep the right speed," he explains, handing me the can. "Too fast and the line is too thin. Too slow and you get drips."

My *K* is clumsy but acceptable.

Halfway through my *E* a passerby, later introducing himself as Alain Auborioux, a former photojournalist with *Le Parisien*, stops to chat. He asks Irvin why he does what he does. Irvin's explanation includes this tale: He and his crew had spent two days painting a massive mural on the side of a bus depot. He noticed two women closely watching them from a window across the street. On the second day, one of the women approached him; Irvin began packing up cans, certain she was going to inform him the police were on the way. But no. The lady explained that her mother—the other woman in the window—had always loved art but was ill and unable to get out to museums. She thanked Irvin for "bringing the art to my mother." Irvin's broader explanation of why he does what he does is simply, "I love to paint."

In her book *Street Fonts*, Claudia Walde explores some of the reasons other graffiti artists do what they do. Wane of New York City posits: "Graffiti is a world of its own. Only writers and artists can understand how it works and they will do anything to keep it alive." TAKECARE of Moscow explains: "Nowadays people don't have much opportunity to shape the reality they live in—their streets, their house, the local shops were all designed and built by others. Graffiti is a chance to take some control back." Sonic of New York City explains he loves the "memories of running the underground subway system in NYC, spray-painting trains until the early morning light, trying not to get killed by passing trains and the electrified third rail. Staying ahead of the vandal squad while making masterpieces. [These are] and always will be the best things about graffiti. It's mine and no one can take it from me."

Ripo explains, "Graffiti usually fills in the negative or forgotten spaces of cities—ignored parts of signs, doors, walls." Mover explains that it "keeps his inner child alive." Cheque of Poland feels it "can arouse emotions in people who just happen to come across it." Faith47, a female artist from South Africa, loves graffiti for "the physical and emotional experiences of exploring cities, the people you meet on the way and the situations you find yourself in." Ashes of Mexico City waxes, "Graffiti is able to do what our leaders can't: bring people together, irrespective of who they are, where they come from, the language they speak, and their religious or sexual preferences."

But for many it's the camaraderie, which is why many artists work in loosely knit crews. Xpome of Bulgaria states, "Through graffiti you meet loads of nice and crazy people who share the same passion and you never have to pay for accommodations again." Dope78 of Germany explains, "The best thing about graffiti is spending a day painting with friends." Like bowling a 300 game, creating a magnificent piece of art is more fulfilling if others are in the alley to help celebrate.[4]

Irvin fits fat caps onto the stems of two spray cans, hands one to me, then begins moving his in a series of horizontal stripes.

"Keep your hand moving and go right up to the outline," he explains. At close range with two aerosol cans going full bore I ask whether he ever wears a mask. He stretches his arms to indicate the abundance of fresh air around us. "Not outside."

As we fill in the letter outlines, Irvin talks about a recent rash of graffiti theft. One well-known Parisian artist—Invader—uses tile mosaics as his medium. He has installed twelve hundred works around Paris and three thousand worldwide. Recently, two men dressed as city workers in yellow vests and armed with extension ladders and chisels removed some of his works—including a well-known representation of the Mona Lisa. On the black market that Mona Lisa could bring hundreds of thousands of dollars.

There can be money in graffiti. Although its roots are countercultural, like garage bands that find themselves playing Madison Square Garden, fame and fortune find a few select graffiti artists. Some consider it "selling out," but the Sirens' call is irresistible.

Keith Haring, who got his start drawing Gumby-like people, babies, and dogs—dancing, interlocked, vibrant, colorful—in New York subways, was soon being commissioned to create murals around the world—at the City of Paris Museum of Modern Art, the Berlin Wall, and the Woodhull Medical and Mental Health Center in New York City. He rubbed shoulders with Madonna, Grace Jones, and Andy Warhol. Though taken by AIDS in 1990 at age thirty-one, his artwork adorned clothing and transistor radios and hung in galleries around the world. An untitled 1982 creation containing iconic dogs, babies, and angels sold for $6,536,000 at Sotheby's in 2017.

Shepard Fairey began his art career decorating skateboards as a teen. Later, as an art student at the Rhode Island School of Design, he created a sticker featuring André the Giant. He began plastering André stickers by the thousands across the East Coast. The stickers became both a pop culture and countercultural symbol. Fairey founded OBEY clothing in 2001; the line includes shirts that combine the best elements of a protest sign and your father's bowling shirt. His net worth is in the tens of millions.⁵

One graffiti artist who has gained fame without crossing to the dark side is London's Banksy—the best-known graffiti artist of today and, perhaps, of all time. He closely guards his anonymity and is equal parts artist, prankster, political activist, and entrepreneur. His *Submerged Phone Boot* (yes, boot) consisting of half a British phone booth surrounded by rubble recently brought over $800,000 at auction. Other works have sold for $2 million.

His 2013 *Better Out Than In* is legendary. For thirty days he lived in New York City and painted, stenciled, and pranked his way across the Big Apple. Included was a pop-up shop in Central Park where original signed canvases sold for $60; few believed they were the real thing—one recently sold for $160,000.⁶ Banksy has also created politically poignant graffiti on the concrete wall separating Palestine and Israel. One image depicts a child floating toward the top of the wall, pulled by helium balloons. As Banksy declares, "Art should comfort the disturbed and disturb the comfortable." He took this last statement to the extreme in 2018 when, within seconds of a painting being auctioned at Sotheby's for $1.4 million, a mechanism built into the frame shredded the picture in front of bidders' unbelieving eyes.

Having a cool name is requisite; there are no John Browns. There are numeric names: 44 Flavours and Pariz One. Impossible to pronounce names: JINSBH, Cxxe, and Vqik. Descriptive names: Panic, Trainboy, and Foggy. Pichiavo is the composite of two artist's names, Pichi and Avo.

Name origins vary: "I was abducted, experimented on, and upon my return, after severe and physical mutilation, took the name Erya51 as homage to my experience." And "I have an appetite for what I do and I like 'y' better than 'i' and didn't feel the

Graffiti aficionado Aurélie Journée points out the wide variety of styles and subject matter found on just one wall in the Belleville area of Paris.

need for 2 'p's, thus apetyte was born." And "People have said my hair looks like a hobo. So, I named myself HOBO."[7]

We continue the SPiKE graffito by reoutlining the letters with more black paint. The "bubble" style we're creating is just one of many. Earlier that morning I'd taken a street art tour with graffiti guide Aurélie Journée. Meeting beneath the Albert Camus statue in a small city park seemed fitting.

During our hour-long walk, we encountered graffiti of every size, shape, race, color, and creed. We came across stenciled artwork of meerkats, Dirty Harry, and Warhol soup cans; and stickers of ani-machines (half-animal, half-machine creatures) and Minions. We viewed paintings on five-gallon bucket lids, security doors, 33 RPM records, trucks, and three-story buildings. We found three-dimensional graffiti made of cat-scratching posts, pianos, and urinals.

We studied graffiti by artists that focus on only alarm clocks, the backsides of people, or cats. We walked down one graffiti-filled street named "*Euh . . .*" When I asked Journée what it means, she explained, "People in France say '*Euh . . .*' if they aren't sure they love something."

As Irvin and I add white accents to the edges of some letters, I continue to keep a paranoid lookout. In 1993, a student from the Singapore American School pled guilty to spray-painting a car. Under Singapore's strict graffiti laws—enacted in 1966 to limit the rash of communist graffiti—he received a fine of $3,500 and a sentence of four months in jail, along with a six-stroke caning. Despite a *New York Times* editorial condemning the sentence and a flood of protests, the caning took place, though the punishment was reduced to four lashes.

Antigraffiti efforts come in a host of approaches.

Instant removal is one oft-used tactic, the theory being that few artists wish to spend hours creating a piece that will be sandblasted or whitewashed into oblivion the next morning. The UK spends $1.5 billion a year on graffiti removal; Germany, $700 million. In 2014, LA spent $7 million cleaning up thirty-two million square feet of graffiti.[8] Graffiti is considered an epidemic in Rome, Paris, Sydney, and London, and removal is a never-ending battle that can backfire. 5Pointz—a complex of a dozen buildings in Queens that provided a two-hundred-thousand-square-foot canvas for gifted graffiti artists from around the world—was whitewashed by its owner one night. Citing the 1990 Visual Artists Rights Act, the artists sued the owner and won a $6.7 million lawsuit.

Preventative coatings are used with some degree of success. One type prevents paint from sticking; another makes paint easier to remove.

Rapid response is another approach. Officials in Australia have begun using "mousetraps"—sensors that detect vapors from spray cans that immediately alert authorities. Modesto, California, recently installed thirty-two surveillance cameras in popular graffiti areas. Citizens can report tagging on the city's GoModesto! mobile app, and its TAG We're On It! Tagging Abatement Program provides do-it-yourself "Tagging Removal Kits." Many large cities offer graffiti hotlines.

Colorful sedimentary layers of paint removed from an often-tagged wall in New York City.

PHOTO BY TOM FENEGA

Fines and jail time always loom, but sentences seem to have limited effect. Anti-graffiti departments in some cities now maintain databases with profiles and styles of graffiti artists. If an artist is apprehended, and there are six other graffiti on file that match the artist's profile, he or she can be convicted on not one, but seven counts.

"If you can't beat 'em join 'em" seems to be gaining traction. Some cities designate spaces where graffiti is allowed—a form of vandalistic inoculation. Rapid City, South Dakota, has its Art Alley; Venice, California, has the Graffiti Pit; Melbourne has Hosier Lane; Zürich, the Rote Fabrik building; and Warsaw, Topiel Street.[9]

We finish by adding a few orange bubbles around "SPiKE." I ask Irvin how many times he thinks this wall has been painted. "Hundreds, maybe thousands, of times," he responds. How long does he think "SPiKE" will remain? "Maybe till tomorrow." He grins.

But that's okay. I've created an ephemeral work of art, albeit a lousy one. A few people will see it . . . maybe some will even pause. I've experienced the camaraderie of working alongside an artist with a story to tell. I spent some time outdoors. I've had the tingly thrill of being a low-grade rebel. I may be a toy, but I get it.

EPILOGUE

Foot Powder, Grass Roots, *and* X-Ray Vision

Nobody made a greater mistake
than he who did nothing
because he could do only a little.
—EDMUND BURKE

WHEW! THAT WAS QUITE THE WALK. GOLD BOND FOOT POWDER
anyone? I wound up with a few blisters, but I finally—finally—got
to find out why my water line froze and where my recycling goes.

I got something else too—something unexpected. I got inspired by a lot of the people we met along the way. I reflect on
Bicycle Bob Silverman, who prodded and pushed Montreal into
being one of the best biking cities in the world. I think about Dan
Buettner and the Blue Zones gang, who've shown entire cities of
people how to live healthier and longer lives. I think about Bea
Johnson, who through her passion and pint jar of trash has changed
the way thousands of us view our garbage. I think about Dr. June
McCarroll in California and Dadarao Bilhore in India—on their
hands and knees—painting center lines and filling potholes, one
by one, to make our roads safer.

These are people so passionate about changing some sliver of the world that they just rolled up their sleeves and dug in. They forged ahead without job title, majority vote, business card, salary, office, or political affiliation. Writer Thomas Friedman refers to these people as "leaders without authority."[1] Where do we find more?

Well, we can start by taking a selfie.

And listening to a pair of voices from the past.

Alexis de Tocqueville—a man absolutely smitten by democracy in America—reminds us that one of the beauties of living in a democracy is that policies aren't decreed from on high by "church and state" but from the bottom up, by "village and congregation."

And anthropologist Margaret Mead expounds, "Never doubt that a small group of thoughtful, committed citizens can change the world; indeed, it's the only thing that ever has."

If I were able to snap my fingers and invite this unlikely duo over for coffee, I believe they'd agree on one thing: grassroots movements are the key to setting change in motion. And if a person has enough passion and the idea has enough power, leaders and institutions at the top take notice. When change is being driven from both top and bottom, it has less distance to travel; change happens faster. People feel empowered.

Sometimes when I'm done walking, I grab a chair, click on the tube, and—as they say in the Midwest—*uffdah!* The news is daunting. Climate change, affordable health care, continents of floating plastic, pandemics, racial inequality, burning rain forests—with problems of such magnitude, of such a national and global scale, it's easy to feel frustrated. But we're not impotent. We can support candidates that understand the significance of climate change, but in the meantime we can turn down the thermostat and plant some trees. We can hope our state legislature bans single-use plastic bags, but for now we can just stop using them. We can cast our votes for better health-care policies, but until that happens, we can bike and hike as our own form of health care. By doing small things, we're not only *becoming part* of something larger, we're helping *create* something larger.

Our actions rub off on one another. When one household on a block installs solar panels, the chances of others doing the same rise substantially. When we see the little bar graphs on our utility bill that show us how our energy usage compares to that of our neighbors, we reduce our electrical consumption. This is peer pressure of the highest order.

As I hammer out the final edits—and very likely as you read this—we find ourselves in the wake of a global pandemic. COVID-19 has opened my eyes to many things; the least of which is how many more people are now walking around the block for exercise, mental health, and, at least some kind of, social interaction. But the pandemic has magnified, and helped me see more clearly, other ideas found on these pages: That we're all in this together—locally, nationally, and globally. That lots of people doing little things— social distancing, wearing masks, taking care of one another—can bend the curve of history in a positive direction. And finally, that many of the people we've met on our walk, those who we oft take for granted, are "essential" in the truest sense. These are the people that keep our water, electricity, and sewage flowing, who pick up our trash, who keep our traffic lights working. I know I appreciate Elizabeth (who already delivers our mail on ninety-five-degree days) and the plow guy (who already clears our streets on Christmas Day) even more. Thank you for your service. All of you.

Researching and writing this book has been a blast. I've ac- quired X-ray vision, able to see through asphalt streets and the walls of water treatment plants. I've also been able to see into people's lives—people I never would have otherwise met. I've found previously mundane things, stoplights, telephone poles, and glasses of water, to be weirdly fascinating. I've learned knowledge is power; and when you know more about how the world works, you make better decisions as you walk through it. I've realized this is *our* block, *our* world, and *our* time to leave a few footprints in the concrete.

Thanks for joining me on this walk.

Where to next?

· ACKNOWLEDGMENTS ·

WRITING IS A CRIMINAL ACT. YOU STEAL TIME FROM FAMILY AND friends, kidnap thoughts from others, kill conversations, torture paragraphs you should have freed long ago. This page is the opportunity to offer reparations, but it'll come up short.

First, endless love and gratitude to Kat—my partner in everything—who supported me and this book selflessly and graciously, who provided precious feedback, and who took care of the zillion details in our lives. Now it's our turn to walk.

Love and thanks to my kids—Tessa, Kellie, Zach, Maggie, and Sarah; to my son-in-laws—Mitch, Charlie, and Mace; to my crew of grandkids—Priya, Louise, Riley, Morgan, Paige, Claire, Anna, Blue, and the one on the way; to my sisters—Patty and Merrilee; to Ma and Pa; and to others who put up with the unruly life of a writer. Thanks to Dan and Gretchen, Jeff and Julie, Paul and Laura, Erik and Kathy, and John and Michelle for friendship, feedback, and support.

Thanks to Laura Dail at the Laura Dail Literary Agency, whose encouragement, energy, and velvet hammer got the proposal into fighting shape and into the right hands. Thank you, Dana Adkins and Deb Phillips, for ten great years and handing off the baton with such grace.

Thanks to the crew at HarperOne—creating a book takes a village (and this village stretches 2,903 miles). To my editor Miles

Doyle for championing this idea and letting me "do my thing"— then making "my thing" way, way, way better. To Anna Paustenbach for carrying the manuscript across the finish line and adding priceless input. To Jessie Dolch and Stephanie Baker for dotting the i's, crossing the t's, and checking the facts. To Courtney Nobile, Allison Ceri, Aly Mostel, Carrie Davidson, and the rest of the marketing and publicity team for banging the pots and pans to get the word out. To production manager Suzanne Quist for dealing with the nuts and bolts. And to creative director Adrian Morgan for the ever-so-clever cover design work.

Much gratitude to the Stillwater Public Library and staff for providing time, space, and resources for research.

Thanks to the crew at Discover Strength for pushing (and pushing) me to be my best. Thanks to the ultimate mechanics, Dr. Paul Lafferty (bones), Dr. Tom Stormont (innards), Dr. Ryan Karlstad (hands), and Dr. Phil Gonzalez (everything else), for getting me back on the road.

This book would not have been possible without those who gave their time, knowledge, and passion to explain the worlds in which they dwell. I hope I translated your words accurately. Any errors are mine and mine alone.

Alleys: Christian Huelsman (Spring in Our Steps, executive director)

Bike Lanes: Gabrielle Anctil (Montreal Biking Advocate, Ghost Bike Project); Robert "Bicycle Bob" Silverman (Le Monde à bicyclette)

Blocks and Neighborhoods: Brent Peterson (Washington County Historical Society); Don Empson (historian, author); Christopher Alexander (University of California, Berkeley, professor emeritus and author of *A Pattern Language*); Roger Tomten (architect and urban planner)

Electricity: Brian Behm (Xcel Energy Allen S. King power plant, director); Pete Schuna (All Tech Electric); Daryl Prihoda (linesman)

Graffiti: Tom Fenega (photographer); Adam Fenega (graffiti aficionado); Irvin (Paris, graffiti artist); Aurélie Journée and Julie (Street Arts Paris)

Lawns: Dr. Paul Koch (University of Wisconsin, Madison, plant pathology); Dr. Doug Soldat (University of Wisconsin, Madison, soil science)

Mail: Elizabeth (Stillwater postal carrier); staff of the Minneapolis Postal Service Processing and Distribution Center

Manholes and Sewage: Matt Simcik (University of Minnesota, School of Public Health); Raymond Smith (Metropolitan Waste Water Treatment Plant); Shawn Sanders (Stillwater, Minnesota, city engineer)

Parking: Donald Shoup (UCLA, professor of urban planning)

Parks: Marcy Breffle (Historic Oakland Foundation, education manager)

Pigeons: Dr. Phil Nelson and members of the Minneapolis Racing Pigeon Club

Recycling: Bill Keegan (Dem-Con, president); Jennifer Potter (Dem-Con, community outreach coordinator)

Roadkill and Litter: Heather Montgomery (author, lecturer, skunk skinner); Bill Jordan (Pocahontas Chamber of Commerce); Ann McLellan (MnDOT, Adopt-A-Highway Program); Dennis Anderson (outdoorsman)

Road Signs and Lines: Josie Tayse (MnDOT state signing engineer); Rick Sunstrom (MnDOT signing supervisor); Ken Johnson (MnDOT, pavement marking and traffic devices engineer); Ethan Peterson (MnDOT, traffic control devices engineer)

Snow: Joe Keding (Shoreview, Minnesota, Public Works Department); Steve Kochendorfer (MnDOT plow driver); Todd Stevens (MnDOT state maintenance engineer)

Squirrels: Dr. Robert Lishak (Auburn University, biological sciences, professor emeritus; John Moriarty (Three Rivers Park District, senior wildlife manager)

Stoplights: Steve Misgen (MnDOT, district traffic engineer); Derrick Lehrke (MnDOT, signal optimization engineer); Willis Lamm (collector, historian)

Street Names: Don Empson (historian, author)

Telephone: Pete Schuna (All Tech Electric)

Trash: Nelson Molina (New York City Department of Sanitation); Chief Keith Mellis (New York City Department of Sanitation); Bea Johnson (Zero Waste Home)

Trees: Mike Branson (Carmel, California, city forester); Guy Carlson (SavATree); Karl Mueller (St. Paul Forestry Division); Gary Johnson (University of Minnesota, forestry professor); Maria Sutherland (Friends of Carmel Forest); Denis Heuer (tree lover)

Walking: Dan Burden, Dan Buettner, Nick Buettner, and the rest of the Blue Zones team (https://www.bluezones.com); Katherine Ball, Jody, and other members of the Monterey Bay Area Walking Club; Linda Lemke (the Nordic Walking Queen)

Water: Stew Thornley (Minnesota Department of Health); Jodi Wallin (St. Paul Regional Water Service); Robert Benson (Stillwater Water Department, manager); Matt Simcik (University of Minnesota, School of Public Health)

A special thanks to Mark Gieseke (MnDOT, Engineering Services Division, assistant director) for connecting me with so many valuable resources.

Finally, thanks to you, dear reader. Writing this book involved finding thousands of loose strands, braiding them into a rope, then handing one end to you and asking you to have faith that the rope is strong enough and long enough to carry you someplace worth going.

Thank you for your trust.

· NOTES ·

Chapter 1: The Front Porch

1. *Stillwater Gazette*, n.d.
2. Duane Johnson, *How a House Works* (Pleasantville, NY: Reader's Digest Association, 1994), 12.
3. Seaside (website), https://seasidefl.com/about.
4. Lynn Freehill-Maye, "American Rediscovers Its Love of the Front Porch," Citylab, November 20, 2017, https://www.citylab.com/life/2017/11/front-porches-are-having-a-moment/546176/.
5. Porch Sitters Union (website), https://porchsittersunion.com/home/.

Chapter 2: Electricity

1. Brad Plummer, "All the World's Power Plants, in One Handy Map," *Washington Post*, December 8, 2012.
2. You can check out the Xcel Energy bird cam at https://birdcam.xcelenergy.com.
3. Maja Beckstrom, "Power Plant Tour Not as Electrifying as Some Kids Might Like," *Pioneer Press*, last updated November 12, 2015, https://www.twincities.com/2010/10/16/power-plant-tour-not-as-electrifying-as-some-kids-might-like/.
4. U.S. Energy Information Administration, "Frequently Asked Questions: What Is U.S. Electricity Generation by Energy Source?," last updated February 27, 2020, https://www.eia.gov/tools/faqs/faq.php?id=427&t=3.
5. Strata, "The Footprint of Energy: Land Use of U.S. Electricity Production," June 2017, https://www.strata.org/pdf/2017/footprints-full.pdf.
6. "How Much Do Wind Turbines Cost?," Windustry, http://www.windustry.org/how_much_do_wind_turbines_cost.
7. Strata, "Footprint of Energy."
8. Strata, "Footprint of Energy."
9. "Power Generation," Xcel Energy, https://www.xcelenergy.com/energy_portfolio/electricity/power_generation.

10. Tony Long, "Jan. 4, 1903: Edison Fries an Elephant to Prove His Point," *Wired*, https://www.wired.com/2008/01/dayintech-0104/.

11. W. Bernard Carlson, *Tesla: Inventor of the Electrical Age* (Princeton, NJ: Princeton University Press, 2013), 368–395.

12. U.S. Energy Information Administration, "Electricity Explained," last updated October 11, 2019, https://www.eia.gov/energyexplained/index.php?page=electricity_delivery.

13. Jennifer Weeks, "U.S. Electrical Grid Undergoes Massive Transition to Connect to Renewables," *Scientific American*, April 28, 2010, https://www.scientificamerican.com/article/what-is-the-smart-grid/.

14. April Mulqueen, "A Natural History of the Wooden Utility Pole," California Public Utilities Commission, July 2017, http://www.cpuc.ca.gov/uploaded Files/CPUC_Public_Website/Content/About_Us/Organization/Divisions /Policy_and_Planning/PPD_Work_Products_(2014_forward)(1)/Utility PoleBook060617.pdf.

15. "Overhead vs. Underground," Xcel Energy Information Sheet, Colorado, May 2014, https://www.xcelenergy.com/staticfiles/xe/Corporate/Corporate %20PDFs/OverheadVsUnderground_FactSheet.pdf.

16. U.S. Energy Information Administration, "Frequently Asked Questions: How Much Energy Does the World Consume by Each Energy End-Use Sector?," last updated September 24, 2019, https://www.eia.gov/tools/faqs/faq.php ?id=447&t=3.

17. Nick Leadmin, "Electrical Safety Statistics," Nickle Electrical Companies, May 27, 2015, https://www.nickleelectrical.com/safety/electrical-safety-statistics.

18. Ronald Wolfe and Russell Moody, "Standard Specifications for Wood Poles," U.S. Department of Agriculture, November 6–7, 1997, https://www.fpl .fs.fed.us/documnts/pdf1997/wolfe97b.pdf; Nelson G. Bingel III, "National Wood Pole Standards," Nelson Research, n.d., https://woodpoles.org/portals /2/documents/WoodPoleCode_Overview.pdf.

Chapter 3: Water

1. "History of Water Supply: Stillwater, Minnesota," https://www.ci.stillwater .mn.us/vertical/sites/%7B5BFEF821-C140-4887-AEB5-99440411EEFD %7D/uploads/%7B63B71F1C-C06A-4305-8FEA-FAFEB436BF0C%7D .PDF.

2. Chris Steller, "Wireless Water Meters a 'Violation of Basic Human Rights,'" Patch, November 14, 2013, https://patch.com/minnesota/stillwater/wireless -water-meters-a-violation-of-basic-human-rights.

3. Rasmus Kerrn-Jespersen, "This Arctic Town Has Running Water for Just Four Months of the Year," ScienceNordic, August 8, 2016, http://science nordic.com/arctic-town-has-running-water-just-four-months-year.

4. Monica Showalter, "Low Sperm Counts," *Breeze* 176 (Winter 2018).

5. "Groundwater Facts," NGWA: The Groundwater Association, https://www .ngwa.org/what-is-groundwater/About-groundwater/groundwater-facts.

6. "International Decade for Action 'Water for Life' 2005–2015,'" United Nations Department of Economic and Social Affairs (website), last updated November 24, 2014, https://www.un.org/waterforlifedecade/scarcity.shtml.

7. David Schaper, "As Infrastructure Crumbles, Trillions of Gallons of Water Lost," NPR, All Things Considered, October 29, 2014, https://www.npr.org/2014/10/29/359875321/as-infrastructure-crumbles-trillions-of-gallons-of-water-lost.

8. Melissa Denchak, "Flint Water Crisis: Everything You Need to Know," NRDC, November 8, 2018, https://www.nrdc.org/stories/flint-water-crisis-everything-you-need-know

9. Quotation attributed to Thomas Fuller.

Chapter 4: Mail

1. The number of P&DCs is in flux, with consolidation creating fewer and larger centers. For instance, in 2011 there were 251 ("Fact Sheet: Processing Facilities," US Postal Service, https://about.usps.com/news/electronic-press-kits/our-future-network/processing_facility_types.pdf).

2. "Where Have All the Blue Boxes Gone?," Save the Post Office, January 2, 2015, https://www.savethepostoffice.com/where-have-all-blue-boxes-gone/.

3. "Security. Law Enforcement. Preserving the Trust," Postal Facts, US Postal Service, https://facts.usps.com/inspection-service/.

4. "MDD Devices (Scanners) Offer Several Safety Features," Postal Times, August 22, 2016, https://www.postaltimes.com/postalnews/mdd-devices-scanners-offer-several-safety-features/.

5. "Common Causes of Postal Worker Injuries," Harris Federal Employee Law Firm, https://www.federaldisability.com/legal-services/federal-disability-retirement/injury-types/common-causes-postal-worker-injuries/.

6. Owen Phillips, "The USPS Is an Extremely Dangerous Place to Work," The Outline, June 30, 2017, https://theoutline.com/post/1836/the-us-postal-service-is-an-extremely-dangerous-place-to-work?zd=1&zi=serk6dhz.

7. Nancy Pope, "Delivering the Hope Diamond," *Smithsonian National Postal Museum* (blog), November 8, 2012, https://www.postalmuseum.si.edu/node/2049.

8. Moriah Gill, "People Used to Mail Their Children Through the US Postal Service, Seriously!," Rare, August 22, 2019, https://rare.us/rare-news/history/people-used-to-mail-their-children/.

9. "Sizing It Up," Postal Facts, US Postal Service, https://facts.usps.com/size-and-scope/.

10. Ruth Alexander and Polly Hope, "Which Country Has the Most Expensive Postal Charges?," BBC News, April 6, 2012, https://www.bbc.com/news/magazine-17614367.

11. Winifred Gallagher, *How the Post Office Created America: A History* (New York: Penguin, 2016), 1.

12. Gallagher, *How the Post Office Created America*, 3.

13. "First-Class Mail Volume Since 1926 (Number of Pieces Mailed, to the Nearest Million)," US Postal Service, About, February 2020, https://about.usps.com/who-we-are/postal-history/first-class-mail-since-1926.htm.

14. "U.S. Post Service Reports Fiscal Year 2018 Results," US Postal Service, About, November 14, 2018, https://about.usps.com/news/national-releases/2018/pr18_093.htm.

15. "The Dead Letter Office, Where U.S. Mail Went to Die," The News Lens, November 4, 2015, https://international.thenewslens.com/article/30102.

16. Laura B. Starr, "Found in Uncle Sam's Mail," *Strand Magazine* 16 (July–December 1898): 148–152.

17. Starr, "Found in Uncle Sam's Mail," 148.

18. Devin Leonard, *Neither Snow Nor Rain: A History of the United States Postal Service* (New York: Grove, 2016), 73.

19. "General Merchandise: (07119-103)," GovDeals: A Liquidity Services Marketplace https://www.govdeals.com/index.cfm?fa=Main.Item&itemid=37158&acctid=4703.

Chapter 5: Telephone Wires and Waves

1. "Justin Bieber Pisses into Restaurant Mop Bucket: 'F*** Bill Clinton!'" TMZ, July 10, 2013, https://www.tmz.com/2013/07/10/justin-bieber-restaurant-mop-bucket-piss-pee-urinate-video-bill-clinton/.

2. "Sports People; Mittleman Sets Record," *New York Times*, May 9, 1986, https://www.nytimes.com/1986/05/09/sports/sports-people-mittleman-sets-record.html.

3. Evan Andrews, "10 Things You May Not Know About the Pony Express," History, last updated August 29, 2018, https://www.history.com/news/10-things-you-may-not-know-about-the-pony-express.

4. Otto Meyer, "American Trenton Breeders," 1980, https://www.racingpigeonmall.com/loft/history/Haffner-Meyers.html.

5. "Telegraphy," Wikipedia, Wikimedia Foundation, last modified March 23, 2020, https://en.wikipedia.org/wiki/Telegraphy.

6. "Tin Can Telephone," Wikipedia, Wikimedia Foundation, last modified March 24, 2020, https://en.wikipedia.org/wiki/Tin_can_telephone.

7. Daniel P. McVeigh, "An Early History of the Telephone 1664–1866," Ocean of Know, http://oceanofk.org/telephone/html/index.html.

8. "Imagining the Internet: A History and Forecast," Elon University School of Communications, https://www.elon.edu/e-web/predictions/150/1870.xhtml.

9. Chris Woodford, "How Cellphones Work," ExplainThatStuff!, last updated May 5, 2019, https://www.explainthatstuff.com/cellphones.html.

10. Tania Teixeira, "Meet Marty Cooper—The Inventor of the Mobile Phone," BBC News, last updated April 23, 2010, http://news.bbc.co.uk/2/hi/programmes/click_online/8639590.stm.

11. S. O'Dea, "Number of Mobile Wireless Cell Sites in the United States from 2000 to 2018," Statista, February 27, 2020, https://www.statista.com

/statistics/185854/monthly-number-of-cell-sites-in-the-united-states-since
-june-1986/.

Chapter 6: Recycling

1. Wisconsin Department of Natural Resources, "Recycling Facts and Figures," last revised December 8, 2016, https://dnr.wi.gov/topic/recycling/facts.html.

2. Laura Parker, "Plastic," *National Geographic*, June 2018, 40.

3. Juliet Lapidos, "Will My Plastic Bag Still Be Here in 2507?," Slate, June 27, 2007, httpoi//olato.com/nowo and politico/2007/06/do plastic bags really-take -500-years-to-break-down-in-a-landfill.html.

4. Claire Thompson, "Paper, Plastic or Reusable?," *Stanford Magazine*, August 29, 2017, https://medium.com/stanford-magazine/paper-plastic-or-reusable-cloth -which-kind-of-bag-should-i-use-c4039575f3f1.

5. Wisconsin Department of Natural Resources, "Recycling."

6. Michael Corkery, "As Costs Skyrocket, More U.S. Cities Stop Recycling," *New York Times*, March 16, 2019.

7. U.S. Environmental Protection Agency, "Facts and Figures About Materials, Waste and Recycling," last updated March 13, 2020, https://www.epa.gov /facts-and-figures-about-materials-waste-and-recycling/national-overview -facts-and-figures-materials.

8. "Recycling Rates Around the World," Planet Aid, September 2, 2015, http:// www.planetaid.org/blog/recycling-rates-around-the-world.

9. Pat Byington, "Recycling Is One of Alabama's Most Important Industries for New Jobs," Bham Now, February 8, 2018, https://bhamnow.com/2018 /02/08/recycling-one-alabamas-important-industries-new-jobs/.

10. Chris Clarke, "5 Cities That Are Recycling Superstars," TakePart, September 17, 2014, http://www.takepart.com/article/2014/09/17/5-cities-are -recycling-superstars.

Chapter 7: Sewers

1. Metropolitan Council Environmental Services, "Decades of Improving Water Quality."

2. Eric Roper, "What Happens After You Flush? The Work Never Stops at St. Paul Plant," *Star Tribune*, December 24, 2015, http://www.startribune .com/what-happens-after-you-flush-the-work-never-stops-at-st-paul-plant /363206821/.

3. Molly Guinness, "1832: The Deadly Epidemic That Helped Shape Today's Paris," RFI, last modified December 2, 2010, http://en.rfi.fr/visiting-france /20101118-1832-epidemic-helped-shape-todays-paris.

4. Suzuki Naoto, "Manhole Cards," Nippon (website), September 8, 2017, https://www.nippon.com/en/views/b06304/?pnum=3.

5. Ed Wodalski, "5 Ways Cities Are Clamping Down on Manhole Cover Theft," Municipal Sewer & Water, February 11, 2015, https://www.mswmag.com

/online_exclusives/2015/02/5_ways_cities_are_clamping_down_on_manhole_cover_theft.

6. Patrick McGeehan, "Manhole Lid Theft Is on the Rise," *New York Times*, May 3, 2012, https://www.nytimes.com/2012/05/04/nyregion/thieves-take-con-eds-manhole-covers.html.

Chapter 8: Trash

1. Edward Humes, *Garbology: Our Dirty Love Affair with Trash* (New York: Avery, 2013), 13. In 2008 the Fresh Kills Landfill began being transformed into a recreational area with sports fields, open space, and trails for hiking, biking, and horseback riding.

2. From the sonnet "The New Colossus" by Emma Lazarus cast onto a bronze plaque mounted inside the pedestal of the Statue of Liberty. Popularized by the 1949 musical *Miss Liberty*.

3. "Municipal Solid Waste Generation, Recycling, and Disposal in the United States: Facts and Figures for 2012," U.S. Environmental Protection Agency, February 2014, https://archive.epa.gov/epawaste/nonhaz/municipal/web/pdf/2012_msw_fs.pdf.

4. Humes, *Garbology*, 271.

5. Humes, *Garbology*, 77.

6. "The Remarkable Evolution of Trash and All Its Dirty Secrets," Trashcans Unlimited, September 23, 2016, https://trashcansunlimited.com/blog/the-remarkable-evolution-of-trash-and-all-its-dirty-secrets/.

7. Cait Etherington, "Excavating the City: A Look at Urban Archaeology in New York," 6sqft (website), February 7, 2017, https://www.6sqft.com/excavating-the-city-a-look-at-urban-archaeology-in-new-york/.

8. "History of the Can: An Interactive Timeline," Can Stats & Info, Can Manufacturers Institute, http://www.cancentral.com/can-stats/history-of-the-can.

9. "The Remarkable History of the Glass Bottle," O.Berk (website), April 8, 2016, https://www.oberk.com/packaging-crash-course/remarkable-history-of-glass.

10. "How Many Plastic Bags Are Used Each Year?," The World Counts, http://www.theworldcounts.com/counters/waste_pollution_facts/plastic_bags_used_per_year.

11. Anne Quito, "A New Swedish Bicycle Is Made from 300 Recycled Nespresso Pods," Quartz, August 13, 2019, https://qz.com/1685111/a-swedish-bike-is-made-from-300-nespresso-coffee-pods/.

12. Dick Sheridan, "Trash Fight: The Long Voyage of New York's Unwanted Garbage Barge," *New York Daily News*, August 14, 2017, https://www.nydailynews.com/new-york/trash-fight-long-voyage-new-york-unwanted-garbage-barge-article-1.812895.

13. Bea Johnson, *Zero Waste Home: The Ultimate Guide to Simplifying Your Life by Reducing Your Waste* (New York: Scribner, 2013), 3.

14. Johnson, *Zero Waste Home*, 39–40.

15. U.S. Environmental Protection Agency, "Facts and Figures About Materials, Waste and Recycling," last updated March 13, 2020, https://www.epa.gov /facts-and-figures-about-materials-waste-and-recycling/national-overview -facts-and-figures-materials.

16. "Components of a Modern Municipal Solid Waste Landfill's Environmental Containment System," Department of Environmental Conservation, https: //www.dec.ny.gov/chemical/23719.html.

17. Sheridan, "Trash Fight."

18. "What Happens to Waste to Energy Incineration Ash?," Eco, https://www .thisiseco.co.uk/news_and_blog/what-happens-to-waste-to-energy-incin eration-ash.html.

19. Julia Pyper, "Does Burning Garbage to Produce Electricity Make Sense?," *Scientific American*, August 26, 2011, https://www.scientificamerican.com /article/does-burning-garbage-to-produce-energy-make-sense/.

20. "Business Insider Today Feature on Pay-As-You-Throw," Pay-as-You-Throw (website), February 14, 2020, http://payasyouthrow.org/2020/02/business-insider-today-feature-on-pay-as-you-throw/.

21. "Helpful Resources," Recology (website), https://www.recology.com/recology -san-francisco/rates/.

22. Katie Brigham, "How San Francisco Sends Less Trash to the Landfill Than Any Other Major U.S. City," CNBC, July 14, 2018, https://www.cnbc .com/2018/07/13/how-san-francisco-became-a-global-leader-in-waste-man agement.html.

23. Michelle Crouch, "24 Things Your Garbage Collector Wants You to Know," Reader's Digest, https://www.rd.com/advice/work-career/garbage-collector -secrets/; Patrick Gillespie, "The $100,000 Job: Garbage Workers," CNN Business, February 25, 2016, https://money.cnn.com/2016/02/24/news /economy/trash-workers-high-pay/index.html.

24. "Refuse and Recyclable Material Collectors: Salary, Job Description, How to Become One, and Quiz," Owl Guru, (website), last updated April 28, 2020, https://www.owlguru.com/career/refuse-and-recyclable-material-collectors/.

25. Jaime Hellman, "Sanitation Gold: NYC Garbage Collection Jobs in Huge Demand," Al Jazeera America, January 13, 2015, http://america.aljazeera .com/watch/shows/real-money-with-alivelshi/articles/2015/1/13/sanitation -gold.html.

26. Robin Nagle, *Picking Up: On the Streets and Behind the Trucks with the Sanitation Workers of New York City* (New York: Farrar, Straus and Giroux, 2013), 16–17.

27. Gillespie, "$100,000 Job."

Chapter 9: Roadkill (and Litter)

1. Heather Montgomery, *Something Rotten: A Fresh Look at Roadkill* (New York: Bloomsbury Children's Books, 2018), x.

2. "How Likely Are You to Have an Animal Collision?," State Farm, Simple Insights, https://www.statefarm.com/simple-insights/auto-and-vehicles/how -likely-are-you-to-have-an-animal-collision.

3. "World's Largest Highway Overpass for Wildlife on Track in California," CBS News, updated August 21, 2019, https://www.cbsnews.com/news /worlds-largest-highway-overpass-for-wildlife-on-track-in-california/.

4. Montgomery, *Something Rotten*, 132.

5. Evan Bush, "Dining on Roadkill: Washington Residents Gather 1,600 Deer, Elk in Law's First Year," *Seattle Times*, July 25, 2017, https://www.seattletimes .com/life/outdoors/roadkill-plan-people-take-1600-deer-elk-off-washington -roads-in-first-year/.

6. "National Eagle Repository, Fact Sheets," U.S. Fish & Wildlife Service, last modified August 10, 2016, https://www.fws.gov/eaglerepository/factsheets .php.

7. Montgomery, *Something Rotten*, 95.

8. Allison Klein, "'Plogging' Is the Swedish Fitness Craze for People Who Want to Save the Planet: It's Making Its Way to the U.S.," *Washington Post*, February 23, 2018, https://www.washingtonpost.com/news/inspired-life/wp/2018/02/23 /plogging-is-the-swedish-fitness-craze-for-people-who-want-to-save-the -planet-its-making-its-way-to-the-u-s/?utm_term=.beade2485dbf.

Chapter 10: Bike Lanes

1. Fred C. Kelly, "The Great Bicycle Craze," *American Heritage* 8, no. 1 (December 1956), https://www.americanheritage.com/great-bicycle-craze.

2. Peter Walker, *How Cycling Can Save the World* (New York: TarcherPerigree, 2017), 66.

3. Joseph Stromberg, "In 1900, Los Angeles Had a Bike Highway—and the US Was a World Leader in Bike Lanes," Vox, June 30, 2015, https://www.vox .com/2015/6/30/8861327/bike-lanes-history.

4. Kelly, "Great Bicycle Craze."

5. Walker, *How Cycling Can Save the World*, x.

6. Winnie Hu, "A Surge in Biking to Avoid Crowded Trains in N.Y.C.," *New York Times*, March 14, 2020, https://www.nytimes.com/2020/03/14/ny region/coronavirus-nyc-bike-commute.html.

7. Walker, *How Cycling Can Save the World*, x.

8. "Bike Culture: Europe vs America," Reliance Foundry, https://www.reliance -foundry.com/blog/biking-usa-europe.

9. "Statistics Library: Protected Bike Lane Statistics," People for Bikes, http: //peopleforbikes.org/our-work/statistics/statistics-category/?cat=protected -bike-lane-statistics.

10. Kay Teschke et al., "Route Infrastructure and the Risk of Injuries to Bicyclists: A Case-Crossover Study," *American Journal of Public Health* 102, no. 12 (December 1, 2012): 2336–2343, https://doi.org/10.2105/AJPH.2012.300 762.

11. Roger Geller, "Four Types of Cyclists," Portland Bureau of Transportation, https://www.portlandoregon.gov/transportation/44597?a=237507.

12. Walker, *How Cycling Can Save the World*, 108.

13. Mark Wagenbuur, "Dutch Cycling Figures," *Bicycle Dutch* (blog), January 2, 2018, https://bicycledutch.wordpress.com/2018/01/02/dutch-cycling-figures/.

14. Walker, *How Cycling Can Save the World*, 67.

15. "Statistics Library."

16. Walker, *How Cycling Can Save the World*, 130.

17. "14 Ways to Make Bike Lanes Better (the Infographic)," People for Bikes, May 15, 2014, https://peopleforbikes.org/blog/14-ways-to-make-bike-lanes-better-the-infographic/; Gemma Alexander, "Are Bikes Lanes Worth the Cost?," Earth911, August 10, 2018, https://earth911.com/business-policy/are-bike-lanes-worth-the-cost/.

18. Alissa Walker, "Bike Lanes Need Barriers, Not Just Paint," Curbed, May 10, 2019, https://www.curbed.com/word-on-the-street/2019/5/10/18527503/bike-lanes-red-cup-project.

19. Walker, *How Cycling Can Save the World*, 111.

20. Walker, *How Cycling Can Save the World*, 160.

21. Walker, *How Cycling Can Save the World*, xiii–xiv.

22. "Bicycle Safety," National Highway Traffic Safety Administration, U.S. Department of Transportation, accessed May 4, 2020, https://www.nhtsa.gov/road-safety/bicycle-safety; Harry Lahrmann et al., "The Effect of a Yellow Bicycle Jacket on Cyclist Accidents," *Safety Science* 108 (October 2018): 209–217, https://doi.org/10.1016/j.ssci.2017.08.001.

Chapter 11: Asphalt Streets

1. "History of Asphalt," National Asphalt Pavement Association, http://www.asphaltpavement.org/index.php?option=com_content&task=view&id=21&Itemid=41.

2. Philip McCouat, "The Life and Death of Mummy Brown," *Journal of Art in Society*, last updated 2019, http://www.artinsociety.com/the-life-and-death-of-mummy-brown.html.

3. Rose Eveleth, "Ground Up Mummies Were Once an Ingredient in Paint," *Smithsonian*, April 2, 2014, https://www.smithsonianmag.com/smart-news/ground-mummies-were-once-ingredient-paint-180950350/.

4. Donald L. Empson, *The Street Where You Live: A Guide to the Street Names of St. Paul* (Minneapolis: University of Minnesota Press, 2006), 204–205.

5. Kate Ascher, *The Works: Anatomy of a City* (New York: Penguin, 2005), 12–13.

6. "Happiness and Cobblestones," *Sunset* 35 (1915): 60.

7. "History of Asphalt."

8. "How Is Asphalt Made?," Quora, accessed June 12, 2020, https://www.quora.com/How-is-asphalt-made.

9. "Market Facts," National Asphalt Pavement Association, http://www.asphaltpavement.org/index.php?option=com_content&view=article&id=891.

10. Ramtin, 2010, question on Engineering.com Q&A, "How Many Years Is the Average Life Time of Standard Asphalt Pavements?," http://www.engineering.com/Ask/tabid/3449/qactid/1/qaqid/4074/Default.aspx.

11. Jay Leone, "Asphalt vs. Concrete Price," HomeSteady, last updated July 21, 2017, https://sciencing.com/asphalt-vs-concrete-price-5622007.html.

12. Plastic Road (website), https://www.plasticroad.eu/en/.

13. "Pothole Damage Costs U.S. Drivers $3 Billion Annually," AAA Oregon, https://www.oregon.aaa.com/2016/02/pothole-damage-costs-u-s-drivers-3 -billion-annually/.

14. Margarita Cambest, "Towson Car Owner Says Pothole Did $2,300 in Damage to Vehicle," *Baltimore Sun*, January 31, 2018, https://www.baltimore sun.com/news/maryland/baltimore-county/towson/ph-tt-potholes-0131 -story.html.

15. "Speed Limit Lowered on Stretch of Baltimore-Washington Parkway Due to Potholes," NBC Washington, March 2, 2019, https://www.nbcwashington .com/news/local/Speed-Limit-Lowered-on-Stretch-of-Baltimore-Wash ington-Parkway-Due-to-Potholes-506598131.html.

16. Brad Tuttle, "Your City Could Pay for Car Damage Caused by Potholes. But It Probably Won't," *TIME*, April 4, 2014, http://time.com/50101/your-city -could-pay-for-car-damage-caused-by-potholes-but-it-probably-wont/.

17. "Potholes and Road Hazard Insurance: What's Covered?," Pothole.info, January 24, 2018, https://www.pothole.info/2018/01/potholes-and-road -hazard-insurance-whats-covered.

18. Pete Barden, "All You Need to Know About Potholes—Facts and Figures, Avoiding Them and Driving Them!," CDG, updated January 2015, http:// www.cdg-cars.com/community/advice/all-you-need-to-know-about -potholes-facts-and-figures-avoiding-them-and-driving-them/.

19. Amit Anand Choudhary, "Pothole Deaths Unacceptable, Hold Authorities Accountable, Says SC," *Times of India*, updated December 7, 2018, https:// timesofindia.indiatimes.com/india/deaths-due-to-potholes-on-roads-not -at-all-acceptable-victim-family-be-given-compensation-sc/articleshow/66 973033.cms.

20. Think Change India, "His Son Lost His Life Because of a Pothole, So He Filled 550 of Them and Launched an App to Report Them," Your Story (website), July 30, 2018, https://yourstory.com/2018/07/mans-son-lost-life -pothole-filled-550.

21. Paving for Pizza (website), https://www.pavingforpizza.com.

Chapter 12: Alleys

1. Karl Quinn, "Laneway Culture Is the Beating Heart of Melbourne. But It Wasn't Always Like This," *Sydney Morning Herald*, March 30, 2017, https:// www.smh.com.au/opinion/out-of-the-way-industrial-a-little-shabby-how -very-melbourne-20170330-gv9z8f.html.

2. Christian Huelsman, "Where the Alleys Have No Name," streets.mn, December 31, 2018, https://streets.mn/2018/12/31/where-the-alleys-have-no-name/.

Chapter 13: Concrete

1. "Concrete Facts," Concrete Helper, http://concretehelper.com/concrete-facts/.

2. Mark Miodownik, *Stuff Matters: Exploring the Marvelous Materials That Shape Our Man-Made World* (New York: Houghton Mifflin, 2013), 71.

3. "Concrete," Wikipedia, Wikimedia Foundation, last modified June 12, 2020, https://en.wikipedia.org/wiki/Concrete.

4. Robert Courland, *Concrete Planet: The Strange and Fascinating Story of the World's Most Common Man-Made Material* (Guilford, CT: Prometheus, 2011), 63.

5. Courland, *Concrete Planet*, 119.

6. Courland, *Concrete Planet*, 185.

7. Courland, *Concrete Planet*, 269–270.

8. "What's the Difference Between Cement and Concrete?," CCA: Concrete Contractors Association of Greater Chicago, https://www.ccagc.org/resources /whats-the-difference-between-cement-and-concrete/.

9. Ayesha Bhatty, "Haiti Devastation Exposes Shoddy Construction," BBC News, last updated January 15, 2010, http://news.bbc.co.uk/2/hi/americas /8460042.stm.

10. Jonathan Watts, "Concrete: The Most Destructive Material on Earth," *Guardian*, February 25, 2019, https://www.theguardian.com/cities/2019 /feb/25/concrete-the-most-destructive-material-on-earth.

11. Anne Beeldens, "An Environmental Friendly Solution for Air Purification and Self-Cleaning Effect: The Application of TIO_2 as Photocatalyst in Concrete," Belgian Road Research Centre https://pdfs.semanticscholar.org /b8b6/e7170836d7648984f8c0af9027f3323a319f.pdf.

12. Courland, *Concrete Planet*, 336–337.

13. "Recycled Concrete," Portland Cement Association, 2010, https://www .cement.org/docs/default-source/th-paving-pdfs/sustainability/recycled -concrete-pca-logo.pdf?sfvrsn=2&sfvrsn=2.

14. "The Effect of Pavement Surfaces on Rolling Resistance and Fuel Efficiency," Minnesota Department of Transportation, Research Services & Library, October 2014, https://www.dot.state.mn.us/research/TS/2014/201429TS.pdf.

Chapter 14: Parking

1. "The Effect of Pavement Surfaces on Rolling Resistance and Fuel Efficiency," Minnesota Department of Transportation, Research Services & Library, October 2014, https://www.dot.state.mn.us/research/TS/2014/201429TS .pdf.

2. Maria St. Louis-Sanchez, "Want to Avoid the Meter Maid? Know Your Odds," Gazette, February 2, 2012, https://gazette.com/news/want-to-avoid -the-meter-maid-know-your-odds/article_f686bfa2-d04b-56fa-9560 -7044dcb5a03e.html.

3. Complus Data Innovations, Inc., "Parking Ticket Statistics of 2018," Passport Labs, https://www.complusdata.com/2019/01/08/parking-ticket-statistics-of -2018/; Leanne, "Top 10 Cities with Highest Revenues from Parking Violations," *Parking Panda* (blog), July 23, 2015, https://www.parkingpanda. com/blog/post/top-10-cities-with-highest-revenues-from-parking-violations.

4. St. Louis-Sanchez, "Want to Avoid the Meter Maid?"

5. Don Reisinger, "Stop Circling: Waze Update Helps You Find a Parking Space," PCMag, September 19, 2016, https://www.pcmag.com/news/347997 /stop-circling-waze-update-helps-you-find-a-parking-spot.

6. Donald Shoup, *Parking and the City* (New York: Routledge, 2018), 22.

7. Michael Kimmelman, "Paved, but Still Alive," *New York Times*, January 6, 2012, https://www.nytimes.com/2012/01/08/arts/design/taking-parking-lots -seriously-as-public-spaces.html.

8. Donald Shoup, "How Donald Shoup Will Find You a Parking Spot," ReasonTV, November 9, 2010, YouTube video, 6:49, https://www.youtube .com/watch?v=uVteHncimV0.

9. Shoup, *Parking and the City*, 2.

10. Monterey Park Municipal Code, "21.22.120 Minimum Parking Spaces Required," http://qcode.us/codes/montereypark/view.php?topic=21-21_22 -iii-21_22_120&frames=on.

11. Mark Schaefer, "Parking Garage Square Footage per Car," Sciencing, August 7, 2017, https://sciencing.com/facts-7576253-parking-square-footage-per-car .html.

12. Donald Shoup, *The High Cost of Free Parking* (New York: Routledge, 2017), 87.

13. Shoup, *Parking and the City*, 9.

14. Kathleen Kelleher, "It's Mine and You Can't Have It," *Los Angeles Times*, May 26, 1997, https://www.latimes.com/archives/la-xpm-1997-05-26-ls -62594-story.html.

Chapter 15: Walking

1. Peter Walker, *How Cycling Can Save the World* (New York: TarcherPerigree, 2017), 47.

2. "Pedestrian Deaths Up Sharply in U.S.," AARP Bulletin, April 2020, 4.

3. Luz Lazo, "Traffic Fatalities Fall; Deaths of Pedestrians, Bicyclists Rise," *Minneapolis Star Tribune*, October 23, 2019, A2.

4. Sally Wadyka, "How to Get the Biggest Benefits of Walking," *Consumer Reports*, last updated November 4, 2019, https://www.consumerreports.org /exercise-fitness/benefits-of-walking/.

5. Kara Mayer Robinson, "Walking," Jump Start WebMD, reviewed May 30, 2018, https://www.webmd.com/fitness-exercise/a-z/walking-workouts.

6. "Walking: Your Steps to Health," Harvard Health Publishing, Harvard Medical School, last updated July 18, 2018, https://www.health.harvard .edu/staying-healthy/walking-your-steps-to-health.

7. Meghan Rabbitt, "11 Biggest Benefits of Walking to Improve Your Health, According to Doctors," *Prevention*, January 29, 2020, https://www .prevention.com/fitness/a20485587/benefits-from-walking-every-day/; Jessica Smith, "10 Amazing Benefits of Walking," *MyFitnessPal* (blog), March 25, 2017, https://blog.myfitnesspal.com/10-amazing-benefits-walking/.

8. "Study: Even a Little Walking May Help You Live Longer," American Cancer Society (website), October 19, 2017, https://www.cancer.org/latest-news /study-even-a-little-walking-may-help-you-live-longer.html; "Research Says Walking This Much Per Week Extends Your Life," Blue Zones (website), accessed June 22, 2020, https://www.bluezones.com/2018/07/research-says -walking-this-much-per-week-extends-your-life/.

9. Rabbitt, "11 Biggest Benefits."

10. Smith, "10 Amazing Benefits."

11. Joseph Mercola, "New Study: Daily Walk Can Add 7 Years to Your Life," Peak Fitness, September 11, 2015, https://fitness.mercola.com/sites/fitness /archive/2015/09/11/daily-walk-benefits.aspx.

12. Mark Fenton, "Walking for Fitness," eMedicineHealth, reviewed on February 6, 2019, https://www.emedicinehealth.com/walking_for_fitness/article_em.htm.

Chapter 16: The Block

1. Jay Walljasper, "What We Can Learn from Europe's Urban Success Stories: How to Fall in Love with Your Hometown," in *Toward the Livable City*, ed. Emilie Buchwald (Minneapolis: Milkweed, 2003), 231–264, http:// jaywalljasper.com/articles/europes-urban-success-stories.html.

2. Christopher Alexander, Sara Ishikawa, and Murray Silverstein, *A Pattern Language: Towns, Buildings, Construction* (New York: Oxford University Press, 1977), 83, and #14, "Identifiable Neighborhood."

3. Alexander et al., *Pattern Language*, #125, "Stair Seats," and #243, "Sitting Wall."

4. Alexander et al., *Pattern Language*, 72, and as discussed in #12, "Community of 7,000."

5. Alexander et al., *Pattern Language*, 118, and as discussed in #21, "Four-Story Limit."

6. Alexander et al., *Pattern Language*, as discussed in #120, "Paths and Goals."

7. Andres Duany, Elizabeth Plater-Zyberk, and Jeff Speck, *Suburban Nation: The Rise of Sprawl and the Decline of the American Dream* (New York: Farrar, Straus and Giroux, 2000), 10–11.

Chapter 17: Pigeons

1. George F. Barrowclough, Joel Cracraft, John Klicka, and Robert M. Zink, "How Many Kinds of Birds Are There and Why Does It Matter?," *PLOS One*, November 23, 2016, https://doi.org/10.1371/journal.pone.0166307.

2. Andrew Blechman, *Pigeons: The Fascinating Saga of the World's Most Revered and Reviled Bird* (New York: Grove, 2006), 9.

3. Jaymi Heimbuch, "18 Most Bizarre Pigeon Breeds," MNN: Mother Nature Network, February 23, 2015, https://www.mnn.com/earth-matters/animals /stories/18-most-bizarre-pigeon-breeds.

4. "Pigeon Facts and Figures," OvoControl, Innolytics LLC, https://www .ovocontrol.com/pigeon-facts-figures/.

5. "Pigeons," PestWorld.org, https://www.pestworld.org/pest-guide/birds/pigeons; Jagjit Singh, "Pigeon Infestation & Health Hazards in Buildings," EBS, October 28, 2014, https://www.ebssurvey.co.uk/news/14/63/Pigeon-Infest ation-Health-Hazards-in-Buildings.html.

6. Alexandra Klausner, "Why Do We Hate Pigeons So Much?," *New York Post*, February 7, 2017, https://nypost.com/2017/02/07/meet-the-mother-of-pigeons/.

7. "Famous People," Pigeon Racing (website), accessed June 12, 2020, http://pigeonracing.homestead.com/FAMOUS_Pigeon_Keepers.html.

8. David Freeman, "Nikola Tesla Fell in Love with a Pigeon—and Six More Freaky Facts About the Iconic Inventor," *Huffington Post*, December 3, 2013, https://www.huffpost.com/entry/nicola-tesla-love-pigeon-facts-inventor _n_4320773.

9. Alexandra Lockett, "The D-Day Messenger Pigeon Reminds Us How Amazing These Animals Are," *Guardian*, November 5, 2012, https://www .theguardian.com/commentisfree/2012/nov/05/d-day-messenger-pigeon -amazing-creatures.

10. Steve Harris, "Feral Pigeon: Flying Rat or Urban Hero?," DiscoverWildlife, https://www.discoverwildlife.com/animal-facts/birds/feral-pigeon-flying-rat -or-urban-hero/.

11. "21 Amazing Facts About Pigeons," Pigeon Control Resource Centre, https://www.pigeoncontrolresourcecentre.org/html/amazing-pigeon-facts .html#intelligent.

12. Lockett, "D-Day Messenger."

13. Mary Blume, "The Hallowed History of the Carrier Pigeon," *New York Times*, January 30, 2004, https://www.nytimes.com/2004/01/30/style/the-hallowed -history-of-the-carrier-pigeon.html.

14. Lily Puckett, "Feed the Birds—While You Still Can," *New Yorker*, June 17, 2019, https://www.newyorker.com/magazine/2019/06/24/feed-the-birds-while -you-still-can?subId1=xid:fr1571166713860dfi.

15. Joel Greenberg, *A Feathered River Across the Sky: The Passenger Pigeon's Flight to Extinction* (New York: Bloomsbury USA, 2014), 5.

16. Greenberg, *Feathered River*, 75.

Chapter 18: Parks

1. "How Do Your City's Parks Stack Up?," The Trust for Public Land, http://parkscore.tpl.org/#sm.0001hjgcmp130zco1tq9ex1q89jni.

2. Jay Walljasper, "What We Can Learn from Europe's Urban Success Stories," Jay Walljasper (website), accessed June 22, 2020, http://jaywalljasper.com /articles/europes-urban-success-stories.html.

3. "Dogs Feeling Wuff in the City Getting a Boost from Prozac," *New York Daily News*, January 11, 2007, https://www.nydailynews.com/news/dogs -feeling-wuff-city-boost-prozac-article-1.262918.

4. Richard Dunham, "Hearst Exclusive: Obama Talks About His Golf Habit, His Favorite Magazines and Personal Frustrations," *Houston Chronicle*, April

10, 2011, https://blog.chron.com/txpotomac/2011/04/hcarst-cxclusive-obama
-talks-about-his-golf-habit-his-favorite-magazines-and-personalfrustrations/.

5. Andy Kim, "Philly's Proposed Green Stormwater Plan," *Governing*, September 2010, https://www.governing.com/proposed-stormwater-plan-phila
delphia-emphasizes-green-infrastructure.html.

6. "Why City Parks Matter," City Parks Alliance, https://cityparksalliance.org
/about-us/why-city-parks-matter/.

7. Stanley Coren, "Want to Make More Friends? Get a Dog," *Psychology Today*, June 24, 2015, https://www.psychologytoday.com/us/blog/canine-corner
/201506/want-make-more-friends-get-dog.

8. Katy Read, "St. Louis Park Tries Out a Parklet," *Minneapolis Star Tribune*, August 2, 2019, https://www.startribune.com/st-louis-park-tries-out-a-park
let/513417202/.

9. Chloe Saraceni, "10 Lovely Public Parklets in San Francisco," *7x7*, July 12, 2019, https://www.7x7.com/10-lovely-public-parklets-san-francisco-26391
77872/rotating-art-outside-luna-rienne.

10. William Bostwick, "Life in the Slow Lane," *Architectural Record*, October 28, 2011, https://www.architecturalrecord.com/articles/2344-life-in-the-slow-lane.

11. Ren and Helen Davis, *Atlanta's Oakland Cemetery: An Illustrated History and Guide* (Atlanta: University of Georgia Press, 2012), 175.

12. Galen Cranz, "Urban Parks of the Past and Future," Project for Public Spaces, December 31, 2008, https://www.pps.org/article/futureparks.

13. Eric Jaffe, "Why Europe's Parks and Playgrounds Are So Much More Active Than America's," Sidewalk Talk, April 20, 2018, https://medium.com
/sidewalk-talk/why-europes-parks-and-playgrounds-are-so-much-more
-active-than-america-s-d1963d569205.

14. Peter Harnik and Ben Welle, "Measuring the Economic Value of a City Park System," The Trust for Public Land, 2009, https://www.tpl.org/sites/default
/files/cloud.tpl.org/pubs/ccpe-econvalueparks-rpt.pdf.

Chapter 19: Lawns

1. "2019 International Creative Lawn Stripes Competition Sponsored by Allett Ltd—Terms and Conditions," Allett (website), https://www.allett.co.uk
/2019-international-creative-lawn-stripes-competition-sponsored-by-allett
-ltd-terms-conditions/.

2. Christopher Ingraham, "Lawns Are a Soul-Crushing Timesuck and Most of Us Would Be Better Off Without Them" *Washington Post*, August 4, 2015, https://www.washingtonpost.com/news/wonk/wp/2015/08/04/lawns-are-a
-soul-crushing-timesuck-and-most-of-us-would-be-better-off-without-them/.

3. Stephen J. Dubner, "How Stupid Is Our Obsession with Lawns? (Ep. 289)," Freakonomics, May 31, 2017, https://freakonomics.com/podcast/how-stupid
-obsession-lawns/.

4. "Cleaner Air: Gas Mower Pollution Facts," People Powered Machines, https://
www.peoplepoweredmachines.com/faq-environment.htm.

5. Bob Morris, "Yard Rage: The Rand Paul Assault," *New York Times*, November 10, 2017, https://www.nytimes.com/2017/11/10/opinion/yard-rage-rand -paul-assault.html.

6. Emily Upton, "Why We Have Grass Lawns," Today I Found Out (website), March 5, 2014, http://www.todayifoundout.com/index.php/2014/03/grass -lawns-2/.

7. "A Day in the Life of Louis XIV," Chateau de Versailles, http://en .chateauversailles.fr/discover/history/key-dates/day-life-louis-xiv#mornings.

8. Hilary Greenbaum and Dana Rubinstein, "Who Made That Lawn Mower?," *New York Times*, March 16, 2012, https://www.nytimes.com/2012/03/18 /magazine/who-made-that-lawn-mower.html.

9. Christopher Borrelli, "How About Rethinking a Cultural Icon? The Front Lawn," *Chicago Tribune*, October 7, 2017, https://www.chicagotribune.com /entertainment/ct-ent-lawns-20171002-story.html.

10. "The Scotts Company History," from *International Directory of Company Histories* 22 (1998), Funding Universe, http://www.fundinguniverse.com /company-histories/the-scotts-company-history/.

11. Ted Steinberg, *American Green: The Obsessive Quest for the Perfect Lawn* (New York: W. W. Norton, 2007), 30.

12. James Mills, "Made the Cut," *The Sun*, September 19, 2018, https://www .thesun.co.uk/news/7292640/gardener-273-hours-pruning-lawn/.

13. "Americans Can Get Pretty Competitive About Lawn Care, Study Finds," SWNS Digital, August 9, 2018, https://www.swnsdigital.com/2018/08 /americans-can-get-pretty-competitive-about-lawn-care-study-finds/.

14. Steinberg, *American Green*, 5.

15. Edith Medina, Jacky Guerrero, and Diego Ramirez, "Water Conservation Efforts: An Evaluation of the 'Cash for Grass' Turf Replacement Rebate Program in Los Angeles City Council District 3," May 6, 2015, https:// innovation.luskin.ucla.edu/wp-content/uploads/2019/03/Water_Con servation_Efforts.pdf.

16. Brad Tuttle, "Front Yard Garden Controversy Revelation: Lawns Are Useless," *TIME*, July 11, 2011, https://business.time.com/2011/07/11/vegetable -garden-controversy-revelation-front-lawns-are-useless/.

Chapter 20: Trees

1. Carmel-by-the-Sea Municipal Code, "Chapter 8.44: Permits for Wearing Certain Shoes," current through November 5, 2019, https://www.code publishing.com/CA/CarmelbytheSea/#!/Carmel08/Carmel0844.html#8.44.

2. "Remarks at the South Dakota Centennial Celebration in Sioux Falls," September 18, 1989, *Public Papers of the Presidents of the United States, George Bush, 1989* (Washington, DC: US GPO, 1990), 1208.

3. Ronald Reagan, speaking before the Western Wood Products Association, San Francisco, March 12, 1966, https://www.snopes.com/fact-check/if -youve-seen-one-tree/.

4. Dan Burden, "22 Benefits of Urban Street Trees," May 2006, https://ucanr .edu/sites/sjcoeh/files/74156.pdf.

5. "Forests Generate Jobs and Income," World Bank, March 16, 2016, http:// www.worldbank.org/en/topic/forests/brief/forests-generate-jobs-and -incomes.

6. Erv Evans, "Tree Facts," NC State University College of Agriculture and Life Sciences, https://www.treesofstrength.org/treefact.htm.

7. Evans, "Tree Facts."

8. Burden, "22 Benefits."

9. Jackie Carroll, "Planting Noise Blockers: Best Plants for Noise Reduction in Landscapes," Gardening Know How, https://www.gardeningknowhow.com /special/spaces/noise-reduction-plants.htm.

10. Carroll, "Planting Noise Blockers."

11. Eric Rutkow, *American Canopy: Trees, Forests, and the Making of Nation* (New York: Scribner, 2012), 6.

12. L. Peter MacDonagh, "How 'Clean and Simple' Becomes 'Dead and Gone,'" Green Infrastructure for Your Community, Deeproot, February 16, 2015, http://www.deeproot.com/blog/blog-entries/how-clean-and-simple-becomes -dead-and-gone.

13. Jill Jones, *Urban Forests: A Natural History of Trees and People in the American Landscape* (New York: Viking, 2016), 48.

14. Jones, *Urban Forests*, 52.

15. "History of the American Chestnut," American Chestnut Foundation, https:// www.acf.org/the-american-chestnut/history-american-chestnut/.

16. Jones, *Urban Forests*, 120–121.

17. Deborah G. McCullough, "Will We Kiss Our Ash Goodbye?," *American Forests*, Winter 2013, https://www.americanforests.org/magazine/article /will-we-kiss-our-ash-goodbye/.

18. Minnesota Department of Natural Resources and the Department of Forest Resources, "Minnesota Certified Tree Inspector Program: Study Guide."

19. Kenneth R. Weiss, "Coastal Panel Rejects Monterey Golf Project," *Los Angeles Times*, June 14, 2007, https://www.latimes.com/travel/la-trw-pebblebeach 14jun14-story.html.

20. "How Many Trees Are There in the World? (Video)," *Scientific American*, September 9, 2015, https://www.scientificamerican.com/article/how-many -trees-are-there-in-the-world-video/.

21. Mark Kinver, "World Is Home to '60,000 Tree Species,'" BBC News, April 5, 2017, https://www.bbc.com/news/science-environment-39492977.

22. E. Gregory McPherson, Natalie van Doorn, and John de Goede, "The State of California's Street Trees," US Department of Agriculture, Pacific Southwest Research Station, April 2015, https://www.fs.fed.us/psw/topics/urban_forestry /documents/20150422CAStreetTrees.pdf.

Chapter 21: Squirrels

1. Richard W. Thorington Jr. and Katie Ferrell, *Squirrels: The Animal Answer Guide* (Baltimore: Johns Hopkins University Press, 2006) 30.

2. Rich Juzwiak, "Crazy Cat People Have Nothing on Crazy Squirrel People," Gawker, April 18, 2013, http://gawker.com/5994975/crazy-cat-people-have -nothing-on-crazy-squirrel-people.

3. "How to Find a Wildlife Rehabilitator," Humane Society, https://www .humanesociety.org/resources/how-find-wildlife-rehabilitator.

4. Bill Adler Jr., *Outwitting Squirrels: 101 Cunning Stratagems to Reduce Dramatically the Egregious Misappropriation of Seed from Your Birdfeeder by Squirrels* (Chicago: Chicago Review Press, 1988), 36.

5. "Seed Dispersal by Wind: How Far Can Seeds Travel?," Arboriculture, January 12, 2016, https://arboriculture.wordpress.com/2016/01/12/seed -dispersal-by-wind-how-far-can-seeds-travel/.

6. Sadie Stein, "Alien Squirrel," *New York Magazine*, February 3, 2014, http:// nymag.com/news/features/squirrels-2014-2/.

7. Thorington and Ferrell, *Squirrels*, 96.

8. Thorington and Ferrell, *Squirrels*, 100.

9. Chris Niskanen, "Southeast Asian Hunters Encourage DNR to Better Manage Minnesota's Squirrel Population," *Pioneer Press*, last updated November 13, 2015, https://www.twincities.com/2009/02/14/southeast -asian-hunters-encourage-dnr-to-better-manage-minnesotas-squirrel-pop ulation/; Louisiana Department of Natural Resources (website), http://www .dnr.louisiana.gov/.

10. Tom Rivers, "'Squirrel Slam' Lawsuit Gets New Life in Court," OrleansHub .com, January 2, 2017, https://orleanshub.com/squirrel-slam-lawsuit-gets -new-life-in-court/.

11. Caroline Davies, "The Ultimate Ethical Meal: A Grey Squirrel," *Guardian*, May 10, 2008, https://www.theguardian.com/lifeandstyle/2008/may/11 /recipes.foodanddrink.

12. Stein, "Alien Squirrel."

13. "History of American Urban Squirrel," ScienceDaily, December 6, 2013, https://www.sciencedaily.com/releases/2013/12/131206132408.htm.

14. John Kelly, "Learn to Speak Squirrel in Four Easy Lessons," *Washington Post*, April 9, 2012, https://www.washingtonpost.com/local/learn-to-speak-squirrel -in-four-easy-lessons/2012/04/09/gIQAV8Jr6S_story.html?noredirect =on&utm_term=.92611f5b03bf.

Chapter 22: Snow

1. "How Streets Are Plowed in Minneapolis," Minneapolismn.gov, last updated December 3, 2018, http://www.minneapolismn.gov/snow/snow_snow-removal -basics.

2. "MN2020: The Incredible Snow Melting Machine," Planet Forward, December 3, 2013, https://www.planetforward.org/idea/mn2020-the-incredible -snow-melting-machine.

3. "What Are the Record 24-Hr Snowfalls by State in the U.S.?," Weather Questions.com, last updated December 15, 2019, http://www.weather questions.com/What_are_the_record_snowfalls_by_state.htm.

4. Andrea Mustain, "The 9 Snowiest Places on Earth," LiveScience, February 7, 2011, https://www.livescience.com/30097-the-snowiest-places-on-earth.html.

5. "Safe Winter Roads Protect Lives," Salt Institute, n.d., http://saltinstitute .org/research/safe-winter-roads-protect-lives/.

6. Heather Black, "Alternatives to Road Salts for Safe Winter Driving," Izaak Walton League of America, December 6, 2017, https://www.iwla.org/blog /clean-water-corner/clean-water-corner/2017/12/06/alternatives-to-road -salts-for-safe-winter-driving.

7. Larry Margasak, "The Blizzard of 1888," National Museum of American History, March 9, 2016, http://americanhistory.si.edu/blog/blizzard-1888.

8. History.com editors, "Great Blizzard of '88 Hits East Coast," This Day in History, March 11, History, last updated March 10, 2020, https://www.history.com /this-day-in-history/great-blizzard-of-88-hits-east-coast.

9. "Idaho Snowplow Driver Dies After Getting Pulled into Blades," *Idaho State Journal*, December 14, 2016, https://www.idahostatejournal.com/news /local/idaho-snowplow-driver-dies-after-getting-pulled-into-blades/article _a7490993-3645-5374-a198-4ff3d367fc99.html.

10. Cristina Flores, "Snowplow Driver Recovering After Plunging Down 300-Foot Embankment," KUTV, January 13, 2017, https://kutv.com/news/local /snowplow-driver-recovering-after-plunging-down-300-foot-embankment.

11. "Flesh Descending in a Shower; An Astounding Phenomenon in Kentucky—Fresh Meat Like Mutton or Venison Falling from a Clear Sky," *New York Times*, March 10, 1876, available at https://en.wikisource.org/wiki/The_New _York_Times/Flesh_Descending_in_a_Shower.

12. Mary Caperton Morton, "Clearing Roadways: A Little Salt Goes a Long Way," Earth, November 17, 2009, https://www.earthmagazine.org/article /clearing-roadways-little-salt-goes-long-way.

13. "Should You Buy a Snow Blower or Hire a Plow Guy?," *Consumer Reports*, December 10, 2014, https://www.consumerreports.org/cro/news/2014/12 /should-you-buy-a-snow-blower-or-hire-a-plow-guy/index.htm.

14. Daniel S. Watson, Brenda J. Shields, and Gary A. Smith, "Snow Shovel–Related Injuries and Medical Emergencies Treated in US EDs, 1990 to 2006," *American Journal of Emergency Medicine* 29, no. 1 (January 1, 2011): 11–17, published March 26, 2010, https://doi.org/10.1016/j.ajem.2009.07.003.

15. Susan Olp, "Avoid a Heart Attack: Shovels Sold in Billings-Area Hardware Stores Warn About Symptoms," *Billings Gazette*, December 12, 2017, https://billings gazette.com/lifestyles/health-med-fit/avoid-a-heart-attack-shovels-sold-in -billings-area-hardware/article_90c307f2-fac1-57c4-bd04-50f5821d647a.html.

Chapter 23: STOP!

1. "Fact Sheet: Improving Safety and Mobility Through Connected Vehicle Technology," U.S. Department of Transportation, National Highway Traffic Safety Administration, n.d., https://www.its.dot.gov/factsheets/pdf/safetypilot _nhtsa_factsheet.pdf.

2. Daniel Patrascu, "Road Traffic History: Before the Streets Got Swamped," AutoEvolution, November 8, 2009, https://www.autoevolution.com/news/road-traffic-history-before-the-streets-got-swamped-12954.html.

3. History.com editors, "Automobile History," History, last updated August 21, 2018, https://www.history.com/topics/inventions/automobiles.

4. "Car Crash Deaths and Rates: Motor-Vehicle Fatality Trends," Historical Fatality Trends, National Safety Council Injury Facts, https://injuryfacts.nsc.org/motor-vehicle/historical-fatality-trends/deaths-and-rates/.

5. "History of Traffic Signal Design, Part Five: The End of an Era," Willis Lamm's Traffic Signal Collection, http://www.kbrhorse.net/signals/history05.html.

6. Rachel Ross, "Who Invented the Traffic Light?," LiveScience, December 16, 2016, https://www.livescience.com/57231-who-invented-the-traffic-light.html.

7. "Street Smarts! Preserving the Past and Saving Money," Willis Lamm's Traffic Signal Collection, http://www.kbrhorse.net/signals/preservation01.html.

8. "Schrader Valve," Wikipedia, Wikimedia Foundation, last modified May 5, 2020, https://en.wikipedia.org/wiki/Schrader_valve.

9. "William Phelps Eno," Eno Center for Transportation, https://www.enotrans.org/about-eno/mission-history/.

10. "Stop Sign–Controlled Intersections: Enhance Signs and Markings—A Winston-Salem Success Story," U.S. Department of Transportation, September 4, 2014, https://safety.fhwa.dot.gov/intersection/conventional/unsignalized/case_studies/fhwasa09010/; Richard A. Retting, Helen B. Weinstein, and Mark G. Solomon, "Analysis of Motor-Vehicle Crashes at Stop Signs in Four US Cities," *Journal of Safety Research* 34, no. 5 (2003): 485–489, https://doi.org/10.1016/j.jsr.2003.05.001.

11. Tim Harlow, *Minneapolis Star Tribune*, February 29, 2020, Section B-4.

12. "Replace Stop Signs That Waste Your Time," CNN Opinion, March 16, 2010, https://www.cnn.com/2010/OPINION/03/16/lauder.new.road.sign/index.html.

13. Clint Pumphrey, "How Roundabouts Work," HowStuffWorks, https://science.howstuffworks.com/engineering/civil/roundabouts1.htm.

14. Associated Press, "Deaths Caused by Drivers Running Red Lights at 10-Year High," NBC News, August 29, 2019, https://www.nbcnews.com/news/us-news/deaths-caused-drivers-running-red-lights-10-year-high-n1047616.

15. "Roundabout Benefits," Washington State Department of Transportation, https://www.wsdot.wa.gov/Safety/roundabouts/benefits.htm.

16. Joel Katz, *From Footpaths to Freeways: A Survey of Roads and Highways in Minnesota* (St. Paul: Minnesota Department of Transportation, 2009), 233–234.

17. "Roundabout," Wikipedia, Wikimedia Foundation, May 31, 2020, https://en.wikipedia.org/wiki/Roundabout.

Chapter 24: Road Lines and Signs

1. Minnesota Department of Transportation, "Signs 101 Course Manual," May 2019, p. 3-2.

2. "Dr. June McCarroll Monument," Atlas Obscura, https://www.atlasobscura.com/places/doctor-june-mccarroll-monument.

3. Jason Torchinsky, "You Owe This Woman Your Life, All of You," July 10, 2012, Jalopnik, https://jalopnik.com/you-owe-this-woman-your-life-all-of-you-5924687.

4. Joel Katz, *From Footpaths to Freeways: A Survey of Roads and Highways in Minnesota* (St. Paul: Minnesota Department of Transportation, 2009), 184.

5. Paul J. Carson, Eun Sug Park, and Carl K. Andersen, "The Benefits of Pavement Marking," U.S. Department of Transportation, Federal Highway Administration, August 1, 2008, https://safety.fhwa.dot.gov/roadway_dept/night_visib/pavement_visib/no090488/.

6. "Rumble Strips and Rumble Stripes: Frequently Asked Questions," U.S. Department of Transportation, Federal Highway Administration, last modified on August 28, 2015, https://safety.fhwa.dot.gov/roadway_dept/pavement/rumble_strips/faqs.cfm.

7. Lyn Jerde, "Columbia County Residents Complain of Rumble Strip Noise," *Portage Daily Register*, February 12, 2018, https://www.wiscnews.com/portagedailyregister/news/local/columbia-county-residents-complain-of-rumble-strip-noise/article_0ec830fe-9f3b-59c0-a8d2-fca1fca682e5.html.

8. Minnesota Department of Transportation, "Signs 101," p. 3-21.

9. "Roman Roads," Wikipedia, Wikimedia Foundation, last modified May 31, 2020, https://en.wikipedia.org/wiki/Roman_roads.

10. Minnesota Department of Transportation, "Signs 101," p. 8-4.

11. Minnesota Department of Transportation, "Signs 101," p. 2-1.

12. Minnesota Department of Transportation, "Signs 101," p. 3-21.

13. Jonah Finkelstein, "The 85th Percentile Speed," Mike on Traffic, http://www.mikeontraffic.com/85th-percentile-speed-explained/.

14. Sebastian Toma, "Speeding Fines in the U.S.—The Worst Places to Go Over the Limit in America," AutoEvolution, November 3, 2016, https://www.autoevolution.com/news/speeding-fines-in-the-us-the-worst-places-to-go-over-the-limit-in-america-112691.html.

15. Daily Mail reporter, "Speeding Swede Clocks up World's Biggest Motoring Fine After 180mph Chase," *Daily Mail*, updated August 13, 2010, https://www.dailymail.co.uk/news/article-1302161/Swedish-driver-gets-worlds-largest-speeding-fine-180mph-chase.html.

16. "What Factors Make You More Likely to Get a Speeding Ticket?," CBS News, September 2, 2017, https://www.cbsnews.com/news/speeding-ticket-data-shows-factors-in-getting-pulled-over/.

17. Mark Macesich, "The Car You Drive May Show Your Odds of Getting a Ticket, Website Says," Santander Consumer USA, August 14, 2017, https://santanderconsumerusa.com/blog/the-car-you-drive-may-show-your-odds-of-getting-a-ticket-website-says.

Chapter 25: Street Names and Numbers

1. Donald L. Empson, *The Street Where You Live: A Guide to the Place Names of St. Paul* (Minneapolis: University of Minnesota Press, 2006), xix.

2. Eric Jaffe, "Where the Streets Have No Names (Only Numbers)," CityLab, March 9, 2016, https://www.citylab.com/transportation/2016/03/mapping -street-numbering-us-new-york-philadelphia/472809/.

3. Empson, *Street Where You Live*, xix.

4. Zillow, "Beyond Main: How Streets Get Their Names," Fox News, updated February 6, 2017, https://www.foxnews.com/real-estate/beyond-main-how -streets-get-their-names.

5. Katy Moeller, "What the Cluck? PETA Finds Name of Rural Road in Idaho Distasteful, Asks for Change," *Idaho Statesman*, July 3, 2019, https://www .idahostatesman.com/news/northwest/idaho/article232238047.html.

6. KION546 News Team, "Carmel Post Office a Place to Get Mail, Meet Friends," KION News, December 2, 2016, https://kion546.com/news/2016 /12/02/carmel-post-office-a-place-to-get-mail-meet-friends/.

7. Empson, *Street Where You Live*, xix.

Chapter 26: Graffiti

1. Lori Zimmer, *The Art of Spray Paint: Inspirations and Techniques from Masters of Aerosol* (Beverly, MA: Quarto, 2017), 6.

2. Bill Davidson, "Verdun—Then and Now," *Yank* magazine, December 1, 1944, 9, http://www.oldmagazinearticles.com/pdf/YANK%20-Verdun%20 1944.pdf.

3. Claudia Walde, *Street Fonts* (London: Thames & Hudson, 2011).

4. Walde, *Street Fonts*.

5. Maximilian Braun, "5 Wealthiest Street Artists Who Know How to Earn Their Living," Widewalls, February 28, 2014, https://www.widewalls.ch /magazine/5-wealthiest-street-artists.

6. Hannah Ellis-Petersen, "Banksy Walled Off Hotel in Palestine to Sell New Works by Elusive Artist," *Guardian*, September 7, 2017, https://www.the guardian.com/artanddesign/2017/sep/07/banksy-walled-off-hotel-palestine -gift-shop.

7. "How Did You Get Your Graffiti Name?," Bombing Science (Graffiti Forums, General Discussion), https://www.bombingscience.com/graffitiforum/threads /how-did-you-get-your-graffiti-name.14727/page-4.

8. Fiona McDonald, *The Popular History of Graffiti: From the Ancient World to the Present* (New York: Skyhorse Publishing, 2013), 118; Aaron Mendelson, "LA Scrubs Away 30 Million Square Feet of Graffiti Each Year," KPCC, September 10, 2015, https://www.scpr.org/news/2015/09/10/54285/la -scrubs-away-30-million-square-feet-of-graffiti/.

9. Juliane Huang, "10 Places Where Graffiti Is Legal," Matador Network, January 29, 2009, https://matadornetwork.com/trips/10-places-where-graffiti -is-legal/.

Epilogue

1. Thomas L. Friedman, "President Trump, Come to Willmar," *New York Times*, May 14, 2019, https://www.nytimes.com/2019/05/14/opinion/trump-willmar -minnesota.html.